SURPRISING
QUANTUM BOUNCES

SURPRISING
QUANTUM BOUNCES

Valery Nesvizhevsky
Institut Laue–Langevin, France

Alexei Voronin
P N Lebedev Physical Institute, Russia

Imperial College Press

Published by

Imperial College Press
57 Shelton Street
Covent Garden
London WC2H 9HE

Distributed by

World Scientific Publishing Co. Pte. Ltd.
5 Toh Tuck Link, Singapore 596224
USA office: 27 Warren Street, Suite 401-402, Hackensack, NJ 07601
UK office: 57 Shelton Street, Covent Garden, London WC2H 9HE

Library of Congress Cataloging-in-Publication Data
Nesvizhevsky, Valery, author.
  Surprising quantum bounces / Valery Nesvizhevsky (Institut Laue-Langevin, France),
Alexei Voronin (P.N. Lebedev Physical Institute, Russia).
    pages cm
  Includes bibliographical references and index.
    ISBN 978-1-78326-595-4 (hardcover : alk. paper) -- ISBN 978-1-78326-596-1 (pbk. : alk. paper)
    1. Ultracold neutrons. 2. Quantum theory. 3. Particles (Nuclear physics). I. Voronin, Alexei,
author. II. Title.
  QC793.5.C643N48 2015
  539.7'213--dc23

                                                            2015005269

British Library Cataloguing-in-Publication Data
A catalogue record for this book is available from the British Library.

Typeset by Stallion Press
Email: enquiries@stallionpress.com

To our parents and families for their love and patience

Preface

There are several *crucial ideas* in the history of knowledge. In their times, they seemed to be absolutely amazing. Their striking effects were due to the fact that they displaced great *illusions*.

The stunning statement that the Earth is a sphere moving through the dark space of the Universe was a real challenge for the imagination of a naive observer; it displaced illusions of the flatness and uniqueness of our world.

Galileo's counterintuitive finding that all objects fall down with an equal acceleration seemed to contradict common sense and everyday experience; however, it proved to be valid and pointed out the *universality of gravity*.

Galileo's law of inertia, which states that nonaccelerated motion of an object occurs without any action from other objects, and thus it could elapse infinitely, displaced a natural illusion that space is a sort of viscous medium, which interacts with moving objects, resists and slows them down. This idea provided the basis for modern physics.

Newton's law of gravitation enhanced the universality of gravity. It assumed that falling apples, moving planets and any objects in the Universe are all governed by the same force; this discovery completed the revolution in minds by definitively displacing the illusion of a net separation between our close surrounding and celestial spheres of distant stars and planets.

Einstein's statement that the speed of interaction between objects can never exceed the speed of light, which is equal in any frame of reference, shook the minds of the early twentieth century; this basic idea of special theory of relativity displaced the illusion of absoluteness of space and time.

However, all these ideas seem easier to accept than a much more striking statement of *quantum* theory, which postulates that everything we believe to be rigid bodies are in fact waves. This is a revelation of an uncompromising revolution.

The reason for the special role of quantum mechanics is that the concept of a rigid body is in the heart of our everyday common sense and all scientific knowledge. Indeed, the idea of *space* itself is based on a rigid body as a ruler. In a world deprived of rigid bodies we lose any confidence.

This lack of confidence is reflected in quantum theory; the only prediction that it allows us to make is a *probability* of a certain outcome of an experiment. The statement on the *wave nature* of everything seems to contradict not only our everyday experience, but also the well-established laws of Newtonian mechanics, which are based on the concept of a rigid body.

The way out proposed by quantum theory is that wave nature is mostly evident for very small microscopic objects, like atoms or electrons; and it is usually masked for larger objects in the same way, just as the spherical shape of the Earth is not easily evident for us when we are looking out through a window.

This is good news, because it saves all preceding achievements of classical (pre-quantum) physics. However, it brings another difficulty. The only way we could perform experiments with tiny objects, which are in fact waves, is via using macroscopic experimental installations, as their wave properties are suppressed due to their large mass. This trick looks like an attempt to play a symphony on a drum.

The contradiction between the classical nature of coordinates and time (which we use as a language of description of our observations) and the wave nature of microscopic objects (which we study) results in a lot of confusion.

That is why we can repeat after Richard Feynman that there is "nobody who really understands quantum mechanics". Nevertheless, the rules of quantum theory prove to be efficient in solving *physical problems*, and predictions of quantum mechanics are extremely concrete and precise.

This book is about quantum phenomena and physical problems. We are going to discuss the wave nature of microscopic objects like neutrons, atoms and antiatoms, which they manifest while bouncing on surfaces in a gravitational field. Such *quantum bouncing* is analogous to *classical bouncing* of an elastic ball, which repeatedly falls down on a floor and bounces up.

Following the rules of quantum mechanics, we could predict the motion of a quantum particle in a gravitational field in the vicinity of a reflecting surface. Due to the relative simplicity of quantum mechanical rules for solving this problem, such predictions are analytically precise. This is a rare case in quantum physics. Such simple and easy-to-understand systems are extremely important.

A famous example of such a fundamental system for electrodynamics is the *hydrogen atom*, the theory of which appeared in the early years of quantum physics. Quantum mechanics predicted that the hydrogen atom can be settled only in states with certain, discrete values of energy. Using this hypothesis Niels Bohr explained an intriguing experimental fact that the hydrogen atom (as well as any other atoms) emits light of only certain frequencies.

Mathematical transparency and self-consistency of quantum description of the hydrogen atom later allowed new physics to be discovered. These new physical phenomena manifested themselves in observations of tiny deviations from earlier predicted values. Even today the hydrogen atom is a powerful reference tool for fundamental precision studies.

Similarly to the hydrogen atom, a bouncing quantum particle in a gravitational field in the vicinity of a mirror can have only certain, discrete values of energy. In other words, it can "raise" in a gravitational field only up to certain "allowed" values of height. These extremely small energy values turn out to be very sensitive probes to the presence of even tiny extra interactions, in particular those between the particle and the mirror.

Quantum bouncing particles could play the role of the "hydrogen atom" in quantum tests of the *equivalence principle*, in explorations of *gravitational properties of antimatter*, in searching for *new fundamental short-range and other interactions beyond the Standard Model* of particle physics, in *surface and thin layer physics, chemistry, structure and dynamics*, and so on.

However, quantum bouncing becomes a promising physical laboratory only if corresponding experiments are precise enough. In this book we will discuss the rich world of *challenging experiments with quantum objects in a gravitational field* and the ways to make our understanding of the quantum world deeper.

We start with apparently simple examples of the bouncing of a classical elastic ball in order to put forward not so simple questions on the origin of motion. In the following chapters we are going to discuss quantum motion of tiny particles in a gravitational field and try to puzzle the reader even more.

We do not assume the reader to have already mastered quantum mechanics or mathematics. Though this book is addressed to students who study quantum mechanics, it is also for anyone who wants to discover or rediscover the mysterious world of quantum physics. In order to make our book easier for understanding, we keep it free of more speculative topics, which are intensively discussed in modern quantum physics and of high interest for us personally, but not yet fully settled.

There is no better way to understand something than solving problems, and thus we supply the book with many of them. Some problems are just useful mathematical exercises, some others require real scientific explorations.

We do not expect the reader to solve all of them, and only hope that one enjoys thinking on them. We provide brief solutions in cases if they are really needed for the smooth development of our further explanations. Otherwise we anticipate writing a separate book with problems and their solutions sometime in the future.

We encourage the reader to move forward without losing inspiration if some pages of our book seem difficult; we suggest that physics is taken as a joy.

Figures in our book are not meant to be faithful and boring illustrations of physics laws. Instead they reflect in an artistic manner some deep basic ideas, which we would like to discuss with the reader.

References at the end of our book are given in chronological order to illustrate the process of accumulation of knowledge; we do not assume that the reader is going to read all these references, but this list could be useful for searching for information on particular questions that have arisen. Some key publications are given in footnotes directly in the text; reading and studying these references could be a useful extra exercise, which the reader could do after finishing the book.

Finally, we would like to highlight one of our many personal motivations for the continued study of physics, one that follows directly from the very nature of scientific research: when you discover a really new result or idea, it often *contradicts common beliefs*. Even more, a really new result has to *strongly* contradict common beliefs. However, in contrast to many other human activities, with physics you have *means and tools to prove the truth*. The truth of physics always surprises us.

In this book we aim to discuss fundamental ideas of physics by means of analyzing a single phenomenon of quantum bouncing, which provides numerous physical realizations. We discuss phenomena, which we believe are surprising, and we try to remove some prevailing illusions.

We are grateful to our colleagues, who share with us the everyday duties, the hopes and joys of scientific research in GRANIT and GBAR collaborations. It is our pleasure to work with you. Many pages of this book appear due to your advice, prolific discussions and disputes. We are grateful to Anna Nesvijevskaya for expressing her artistic vision of the physical concepts in the figures of our book.

Contents

Chapter 1

The Fast, the Heavy and the Free

1.1 Classical Equivalence Principle

"All truths are easy to understand once they are discovered; the point is to discover them."[1]

Galileo Galilei

Anyone, who has ever watched a falling apple, probably noticed that apples are more dangerous than leaves. This is so in particular because of the different velocity they gain during the fall. A hailstone falls faster than a light water drop, which in a turn falls faster than an even lighter snowflake (the difference in the gained velocity is a good reason why people are likely to prefer a snowstorm to a hail fall). Aristotle would not be surprised to observe such a regularity![2]

These simple examples are manifestations and proofs of the well-established practical fact that the acceleration of a falling object varies as a function of its mass and shape.

However, as soon as *free fall* is concerned, this obvious statement turns out to become an illusion.

A famous legend[3] describes Galileo's experiments in Pisa exploring classical free fall. According to this legend, Galileo Galilei performed a spec-

---

[1]Galilei, G. (1632). *Dialogue Concerning the Two Chief World Systems*, translated by Stillman Drake, University of California Press (1953, revised 1967).

[2]Aristotle supposed that the speed at which two identically shaped objects sink or fall is directly proportional to their weights and inversely proportional to the density of the medium through which they move.

[3]Viviani, V. (1654). *Racconto Istorico della Vita del Sig.r Galileo Galilei*, Accademico Linceo, Nobil Fiorentino, Primo Filosofo e Matematico dell'Altezze Ser.me di Toscana.

tacular demonstration that objects with different mass fall down identically from the top of the Leaning Tower of Pisa ("Torre pendente di Pisa") when dropped simultaneously. The qualitative statement that gravity accelerates all objects in an universal manner, independently of their mass and composition, contradicted both the traditional doctrine, common beliefs and everyday experience.

What is the actual reason for this striking contradiction between seemingly so similar observations of falling objects performed by different people?

In contrast to naively observing the falling and gliding leaves in the forest on a windy or calm day, or heavy stones falling from the top of a rock in the mountains and breaking up on its foothills, Galileo Galilei succeeded to step beyond the seemingly simple reality and to distinguish *false effects* (like, for instance, the effect of air resistance) and the *principal phenomenon* of unrestricted free fall in order to arrive at his nontrivial and revolutionary conclusion.

Some people call Galileo Galilei "the father of science". We are not going to contribute to the disputable field of priorities in science. What is clear, however, is that he successfully put the exploration of nature on a ground of solid *scientific proofs*, based on the *concept of physical experiment*. This concept is crucial for modern scientific knowledge, and we are going to devote particular attention to its important features, as well as the world of challenging physical experiments.

In particular, the difficult art of eliminating, suppressing or at least accounting for false effects is the essence of any exploration of nature, and a key feature of a properly designed physical experiment. One unrecognized false effect would spoil thousands of correct measurements and would lead to unreliable conclusions concerning the whole study. Even if it is as simple as observations of freely falling objects.

In contrast to earlier times, today it could seem obvious for a modern reader that a false effect like the air resistance effect is only an unnecessary complication, which disturbs the "true" free fall phenomenon. However, we invite the reader to ignore textbooks for a moment and to assume that viscosity is an intrinsic property of our world, so that any motion, which started at some moment, would be slowed down by properties of the space itself. Doesn't this hypothesis look reasonable?

Thus, Galileo's conclusion that air resistance is a false effect is in fact a deep statement about the properties of space itself. It states that the background of everything happening in our world is *empty space*, which does not interact with material objects.

This is not a trivial conclusion, because our everyday experience suggests something opposite; it prompts that space around us is always full of matter and that "nature abhors a vacuum" (following a translation from Aristotle). However, it is an extremely fruitful conclusion because it allows for *quantitative* understanding of motion.

(*Note*: Curiously, 400 years later, the question on whether vacuum is indeed empty, and what vacuum in fact is, induces a lot of serious discussions at a new qualitative level.)

Another universal difficulty, which awaits us on the way to understanding nature, is an always limited *accuracy of measurements*, and thus another key feature of a properly designed physical experiment is sufficiently high and always increasing *precision*.

The same legend about the Pisa tower tells us that Galileo Galilei had to count his own pulse in order to measure short fall-time intervals. A quantitative expression, which relates the height H and the time t of fall, was established later by Christian Huygens.

In order to achieve his result, Christian Huygens radically improved the precision of mechanical clock.

Though the expression for the time of free fall is widely known, it is worth reproducing it:

$$t = \sqrt{2H/g}. \tag{1.1}$$

There is a physical discovery encrypted in this expression. It is encoded in the symbol g. We emphasize that g here has the meaning of the *strength of gravity* near the surface of the Earth. The deep sense of this expression is not in performing a mathematical exercise of calculating a time of motion at a constant acceleration g provided that a height H is given. It consists of establishing a *relation* between *space* (H), *time* (t) and the *gravity* strength (g).

We are going to return to a general discussion of space, time and gravity below. First, however, we invite the reader to feel the taste of experimental exploration of fundamental properties of nature. In the following few pages we are going to discuss how a simple study of free fall could become a source of deep physical questions.

Let us imagine how one could perform a Galileo-type experiment without making a trip to Pisa, and even without leaving your room. Such a simple experiment might be more *precise* and free of major *false effects* than the storied Galileo's measurement! It could be carried out in the limited space of your room or outside it, using an elastic bouncy ball that falls from a given height and bounces back after a short interaction with a surface of a floor or a table.

This simple experiment would be an important source of inspiration for our further discussion.

We propose the reader to perform an experiment very much like an experiment with a bouncy ball shown in figure [1.1], but carried out with particular care. A bouncy ball of good quality would keep moving in the gravitational field of the Earth for a considerably longer time than the time of fall of a heavy object from the top of the Leaning Tower of Pisa.

In order to provide a high level of reliability and accuracy in your experiment, choose a ball that interacts with the surface for a much *shorter time* than the time of free fall between two consequent bounces. Besides, the bouncier the ball, the better it is for our purpose, as it would continue bouncing for a *longer time* until significant dissipation of its energy.

A couple of important points should be taken into account:

i) An apple of any sort depicted in the frame in figure [1.1] would fall only once, so it would not be the right choice... We are going to use a long period of bouncing as a major benefit of our approach.

ii) The ball in figure [1.1] is strongly deformed at the moment of its interaction with the floor. Thus the time of its interaction with the floor is not negligible compared to the time of its rise and fall in the gravitational field of the Earth. Therefore, this ball is rather sensitive to the peculiarities of this interaction, which is not of interest for us at the moment, and thus it is less suitable for studies of free fall, which we would like to focus on. An evident conclusion: avoid such a deformation to a maximum extent.

In contrast to the investigation of free fall of objects from the Pisa tower, a convenient property of our proposed experiment with bouncy balls consists of the fact that bounces are repeated *periodically* many times, and thus we can measure this period precisely after several consequent bounces due to at least two factors: an increase in the total *observation time* and a better *control of systematic* effects. Thus the accuracy would be improving simultaneously with the increase of the total time of bouncing.

The motion of the ball in figure [1.1] is illustrated in an intuitively evident manner, like a sequence of photos made with equal intervals. Thus the distance between two neighboring positions of the ball decreases as a function of height. The velocity of a classical object at its turning height in

Fig. 1.1   Higher does not mean longer.

the gravitational field even equals zero! The intuitively evident conclusion from such a presentation is that a ball spends most of its time in the top part of its trajectory.

So do not make too much effort to increase the initial bouncing height: you do not gain a lot this way, as the measurements would become more complicated and many systematic false effects would significantly increase.

The ball in figure [1.1] deviates significantly from the perfectly vertical direction at the moment of its interaction with the floor. This deviation might result from asymmetries of the ball or the floor in terms of the density distribution inside the ball or in terms of an asymmetric distribution of elastic properties of the ball or the floor. It might also result from rotations of the ball or from roughness of the floor and thus from mixing of motions of different kinds. Avoid such false effects to maximum extent.

Let us establish the law of free fall. Imagine that you are not yet aware of the existence of relation (1.1). You are only trying to guess about a proper relation between parameters of falling down of a ball in the gravitational field. In order to measure a value of the dropping height and a value of the falling time we would need a clock and a ruler. Then we could drop gently a bouncy ball from two different heights $H_{1,2}$ with zero initial velocity and measure two corresponding periods $T_{1,2}$ of its bouncing, so that:

$$\frac{T_1^2}{T_2^2} = \frac{H_1}{H_2}. \tag{1.2}$$

With a bouncy ball, you could also solve a less evident problem of measuring time and height even without having a clock and a ruler.

Problem 1.1. *Propose an experimental method to establish, without using a ruler and a clock, the above law (1.2): $\frac{T_1^2}{T_2^2} = \frac{H_1}{H_2}$, which relates the periods of bouncing ($T_{1,2}$) and the heights of bouncing ($H_{1,2}$) of an ideal bouncy ball.*

Note: When thinking about a scheme of such an experiment, assume that you have got such an ideal bouncy ball that its motion is defined at any moment only by its free fall in the gravitational field of the Earth, while the time of its interaction with the floor is infinitely small and the dissipation of its energy due to its interaction with the floor as well as due to the air resistance is negligible.

Note: You do not need an (absolute) ruler for a relative measurement.

After simply looking to the above expression (1.2), one could conclude that the time t of free fall, as well as the period T of bouncing, are proportional to the square root of the initial height $\sqrt{H}$ of free fall or bouncing, with some coefficient. (*Note:* thus we have nearly got the equation (1.1) with the only unsolved problem related to the value of a coefficient in it.) Is this coefficient the same for various cases? Does its value change as a function of the value of mass or other properties of the ball?

There is no immediate answer to these questions hidden in equation (1.2) itself. In order to clarify this point, we are going to perform an experimental investigation with bouncy balls.

Problem 1.2. *Using a bouncing ball, confirm experimentally the law of free fall (1.1), which relates the time of free fall (t), the height of free fall (H) and the acceleration of free fall (g): $t = \sqrt{2H/g}$.*

Note: As long as any residual systematic effects are present (in particular those associated with the air resistance, with a finite time of interaction of the ball with the floor and also with the energy dissipation due to the interaction of the ball with the floor), the above equality could be proven experimentally only with a finite accuracy. Ignore all small related corrections for a moment.

Note: In a properly designed and performed experiment with a good bouncy ball, any residual false effect should give only small corrections to the principal phenomenon.

Exercise 1.1. *Confirm experimentally that the law of free fall (1.1) $t = \sqrt{2H/g}$ stays the same for any value of the bouncing-ball mass.*

Exercise 1.2. *Using a bouncing ball, a clock and a ruler, measure the value of the free fall acceleration (g).*

Problem 1.3. *Evaluate the relative accuracy associated with the mentioned residual systematic effects (the air resistance, the finite time of interaction of the ball with the floor, the energy dissipation), which could be achieved in your measurements in the best case.*

Note: for each systematic effect, search for an independent method for its evaluation.

As soon as you repeat your experiment with several balls of different mass, which are dropped simultaneously from an equal height, you could rapidly arrive at a *hypothesis* that this coefficient is universal for all objects independent of the value of their mass. It is worth mentioning that this hypothesis is a starting point to the *general theory of relativity.* (A good grasp of this theory might help you in case your neighbors are unhappy with the periodic knocking of a ball on their ceiling.)

Exercise 1.3. *Find a compact analytical expression for the vertical position of an ideal (that means here: with no energy dissipation and with infinitely short time of interaction of the ball with the floor) bouncing ball at any moment of time.*

Calculate the mean height of the ball and its mean velocity during half a bouncing period.

The motion of two such bouncy balls would be particularly beautiful, if the ratio of their initial dropping heights H_1/H_2 is equal to the square of the ratio of two even numbers. If it is so, then after a certain number of periods, the bouncy balls would hit the surface simultaneously. Such a coincidental event could be identified with a high temporal resolution, even aurally. This circumstance could be used for testing the law (1.1, 1.2) with higher accuracy. Thus you have a kind of precision clock, which could be used as a tool for *applications.*

Consider an eventual application. Assume, for instance, that the law of free fall is slightly violated because of a hypothetical or actual reason (one could think of an additional interaction of a new type unknown in physics, or this violation is just due to a small known external perturbation, which you have introduced on purpose). We would like either to measure a degree of violation or to constrain it with a sensitivity imposed by the accuracy of your experiment.

(*Note:* In fact, this scenario is typical for experiments searching for interactions of a new type in modern physics.)

Violation imposes that the period T_1 is no longer an integer multiple of the period T_2 for the two balls with a specially chosen ratio of initial dropping heights: $H_1/H_2 = n^2$. Instead the period of the first ball relative to the period of the second ball is given in the expression $T_1 = nT_2 + \Delta$, where the delay Δ is a measure of violation of equation (1.2). If the violation is relatively small, i.e. if $\Delta \ll T_2$, then after a few periods of T_1 both balls would still touch the surface nearly simultaneously, and the small value Δ would be difficult to identify.

However, after a significant number k of bouncing periods, i.e. after the total time kT_1, the difference between the moments of landing of the two balls would be accumulated thus reaching the value $k\Delta$; this accumulated time difference would be k times larger than the single difference Δ observed after one period of T_1. Thus, small time differences, which arise due to slight violations, could be observed more easily if the balls keep bouncing for a long enough time.

The latter statement is valid, as a general rule, if you are measuring the absolute time needed for k bounces of each ball. However, the accumulated difference could be revealed in a much more spectacular manner if you compare it with the period of one bounce, or rather you compare the phases of landing of the two balls after k bounces. In order to verify that your *instrument* is well suited for further experiments, perform the following exercise.

Exercise 1.4. *Take two equivalent bouncy balls and drop them simultaneously from the same height to a rigid floor. Avoid initial rotations and accelerations of the balls.*

Check that the balls would keep bouncing almost synchronously (with the precision imposed by the equality of their initial conditions, their properties and external perturbations).

Measure the time needed for the difference between the instants of landing of these two balls to become distinguishable aurally.

Exercise 1.5. *Introduce on purpose a slight difference in the motion of the two balls in the previous exercise, for instance, by means of placing on the floor under one of the bouncy balls a large piece of thick paper, which affects the elastic properties of the ball–floor collision.*

Observe how the difference of the bouncy ball landings accumulate as a function of the total bouncing time.

Use this method for comparing and characterizing qualitatively different floors (kinds of paper).

Hopefully, after solving all these simple problems and after performing these exercises, you have already appreciated the power of the method of bouncing balls. Indeed, instead of dropping various massive objects from the top of the Pisa tower, one could rather observe falling objects that spend a significantly longer time bouncing back and forth on a floor or on a table, and thus one could reveal not only general characteristics of free fall but also tiny differences between the balls' motions.

This approach underlines the key idea that initially synchronized oscillations could manifest spectacular deviations from their correlated motion, under an even *small perturbation* provided this perturbation acts for a long enough time.

Thus, high-quality bouncing balls could also be used as very sensitive *instruments*.

(*Note*: In more practical terms, if your mechanical watch is late, this might mean that it measures very tiny effects of friction. Such extremely small effects are accumulated in your watch and become easily visible after a sufficiently long time.)

In particular, this idea plays an important role in modern precision studies of fundamental laws of nature, based on *quantum interference*, which we are going to discuss below.

Problem 1.4.   *Two bouncy balls are dropped simultaneously from the heights H_1 and H_2 onto a drum.*

How does the time interval between two successive hits evolve as a function of the total bouncing time?

Could both bouncing balls be simultaneously found in their initial positions at some time instant t?

Note: Neglect the dissipation of energy of the balls.

Problem 1.5. *A lot of tiny bouncy balls are placed initially between a height H_{min} and a height H_{max} with some known density distribution $\rho_0(z)$; then they are dropped simultaneously.*

How would their density distribution $\rho(z)$ evolve as a function of the total bouncing time?

In order to get an idea concerning the actual (absolute and relative) precision, which could be achieved with bouncy balls, let us study false effects in our free fall measurements (like Galileo Galilei did in his investigations). One could state that an interaction of a ball with a surface (even if it is brief) is an evident source of false effects and thus it is a source of deviations from the free fall law (it is sufficient only to look to figure [1.1] to appreciate the importance of this problem).

However, we could reduce the influence of an (a priori poorly known) interaction of a ball with a surface not only by choosing the most elastic balls and floors, but also by proper accounting of the residual systematic effects.

To start with, we could reasonably assume that the time of interaction of the ball with the surface does not change (or slightly changes) as a function of the height H of bounce. This assumption is valid at least if the interaction of the ball with the surface could be approximated as the successive compression and decompression of an elastic element of a harmonic oscillator. (*Note*: It is known that the period of a harmonic oscillator does not change as a function of its amplitude.)

If the deformation is small enough, then compression/decompression could be thought of as harmonic oscillations with a high accuracy. In this case, and also taking into account equation (1.1), the period of the bouncy ball could be calculated as follows:

$$T = 2\sqrt{2H/g} + \delta T, \qquad (1.3)$$

where δT is the time of interaction of the ball with the surface (to be evaluated or compensated for).

As far as it is proven independently in one or another way that the time δT of interaction of the ball with the surface does not change as a function of the initial dropping height H of the ball (this condition is important), the time of interaction cancels out in the difference $(T(H_2) - T(H_1))$ of two bouncing periods, which are measured for two different initial dropping heights, or equivalently it cancels out from the expression for the following derivative:

$$\frac{dT}{dH} = \sqrt{\frac{2}{gH}}. \tag{1.4}$$

The statement that the time of interaction δT does not change as a function of the initial dropping height H of the ball could also be tested with the bouncing ball itself. However, the accuracy of such a test would be compromised by simultaneous action of another false effect associated with the air resistance.

Thus the expression (1.4) could be tested experimentally by means of comparing the results of a set of measurements with dropping a bouncy ball from different initial heights while keeping all other experimental conditions equivalent to a maximum accuracy.

This expression (1.4) does not contain explicitly the time δT of inter-action of the ball with the surface anymore. Thus the proper choice of a physical value to measure could help us in eliminating false effects, and in considerably reducing systematic uncertainties. We would like to note that the expression (1.4) is valid even for a bouncy ball with energy dissi-pation as long as our assumption on the independence of the ball–surface interaction time on the fall height is valid.

In order to clarify the question on how, in practical terms, we could use a ball even with energy dissipation for precision measurements, let us solve the following problem.

Problem 1.6.   *A bouncy ball is dropped from the initial height H. As-sume that it loses a small fraction Q of its total energy in each collision with a floor (so that its total energy after n-th collision E_n becomes equal $E_{n+1} = (1 - Q)E_n$).*

How long could the ball continue bouncing until the moment when its total energy would decrease by a factor of two? or by a factor of thousand?

How many times would it bounce until that moment? How much energy does the ball lose during the first second, during the first 10 seconds, during the first 100 seconds? Explain this sequence qualitatively.

How does the height h_n of n-th raise change as a function of the bounce number n?

Now we are better prepared to compare the accuracy and reliability of experiments of two kinds: the dropping of objects from the top of the Pisa tower (which you see through the open door in figure [1.1] on one hand, and playing with elastic balls bouncing on a rigid floor (the experiment which you also see in figure [1.1]) on the other hand. (Appreciate that the *inclined* Pisa tower is a good choice when you are going to study the fall of objects from a large height!)

To define the meaning of this comparison, we should mention that in a typical experiment, which is free of major systematic effects, the accuracy increases proportionally to the *observation time*; also the systematic uncertainties decrease proportionally to the values of *systematical corrections* to be introduced. In a properly designed experiment, you would like to improve both these factors simultaneously and also to avoid any extra disturbing phenomena to a maximum extent.

Let us mention that the height of the famous Leaning Tower of Pisa is about 56 meters. The time of the fall of an object from the top of this tower (estimated without accounting for the effect of air resistance) is as short as 3.4 seconds. This time interval is significantly shorter than the total time of bouncing of an elastic ball of good quality, dropped from the height of an adult on a stone floor; the bouncing time is typically about 30 seconds, and it could be even further increased under certain conditions.

The height of a corresponding tower, which would provide an equal 30 seconds of free fall (also estimated without accounting for the effect of air resistance) is as much as 4400 meters! You would not find easily such a free fall distance available for experiments in Galileo's time, and you would not easily do a proper measurement in those times in the absence of modern means of communication between the top and the bottom of such a tower.

Besides, there is a *systematic reduction* of the air resistance false effect associated with the bouncing ball method. It causes only a *small correction* to the free fall law in the case of a bouncing ball, while it is the dominant false effect in the case of an object, dropped from a high tower. The physical reason for the systematic difference in the contributions of the air resistance effect consists of considerably different mean velocities achieved by the objects in these two experiments.

As the air resistance (the air viscosity) decreases with decreasing the value of velocity, bouncy balls get a considerable advantage over simply falling objects. In the opposite case of relatively large velocities (like those corresponding to the fall of an object from the top of a 4400-meter tower) the force of air resistance could be as large as the gravitational force; then the velocity would rapidly achieve its saturation value and would not increase further.

Thus the experiment would lose any sense in such a limiting case as we would then start "investigating" a different physical phenomenon of the air viscosity.

In order to get some practical feeling concerning the values of eventual contributions of the mentioned effect of the air resistance, perform the following exercise.

Exercise 1.6. *Calculate the value of the constant (saturation) velocity, which a ball with the radius r and the density ρ_r would achieve while falling from a tall tower.*

Note: Take into account that the air resistance force for a spherical object with the radius of a few cm and the velocity larger than a few meters per second could be expressed as follows: $F = 0.22\pi r^2 v^2 \rho$, where $\rho \simeq 1.29$ kgm^{-3} is the air density.

The advantage of using relatively small bouncing balls dropped from relatively small heights, compared to dropping larger objects from larger heights, consists of not only a considerable reduction in the air resistance effect, but also of a possibility to *account for* the residual small effects in framework of a rigorous mathematical procedure. A reader acquainted with analytical calculus could establish this procedure by performing the following exercise.

Exercise 1.7. *Calculate the change in the period of a bouncy ball due to counting for the effect of air resistance. Study a case of a small bouncy ball (a few times a fraction of millimeter), which is dropped from the height of a few centimeters.*

Note: In this case the air resistance is proportional to the velocity of the object according to the following law: $F = 6\pi r \mu v$, where $\mu \simeq 1.7 \cdot 10^{-5}$ $kgm^{-1}s^{-1}$ is the air dynamical viscosity.

We could conclude that a major benefit of a bouncy ball experiment, compared with an experiment that involves dropping heavy objects from the top of the Pisa tower, consists of a much longer total time of motion of your probe in the gravitational field of the Earth, or in other words it consists of a much longer observation time.

Why did Galileo Galilei perform a different experiment? We are sure that he thought about all the problems described above: the total observation time to be increased, the air resistance effect to be reduced, the dissipation of energy because of the interaction with the floor to be suppressed...

And he found an alternative solution.

In fact, even if Galileo Galilei was dropping objects from the top of the Pisa tower for illustrative purposes, his actual experiment was different. He used balls rolling over inclined planes with precisely the same purposes: to increase the effective time of fall in the gravitational field of the Earth, to decrease the effect of air resistance due to smaller achieved velocities, to decrease the effect of the interaction with the floor due to rolling of the balls...[4]

We hope the reader has got a feeling for the experimental study of nature with this small treatise of a seemingly simple phenomenon.

In the following chapters we are going to show that the total time of observation is a crucial parameter not only for using classical bouncy balls in precision or sensitive measurements but also for precision or sensitive studies with quantum particles, like those with neutrons, atoms and antiatoms, which could bounce (in another, quantum, manner) in the gravitational field of the Earth on a material surface.

We are going to see that the effect of air resistance is absent there, and the interaction of a quantum particle with a floor is very close to the idealized case of reflection from a perfect mirror.

We are going to see also that the accuracy achievable with quantum bouncing of neutrons, atoms and antiatoms could largely exceed the accuracy of classical measurements with bouncy balls, as described here.

Study 1. *Investigate what relative accuracy for the value of the gravitational acceleration g you could achieve.*

---

[4]Sharratt, M. (1996). *Galileo: Decisive Innovator*, Cambridge University Press, Cambridge, p. 75.

Try to account for the energy loss within a certain model.

Analyze at what level of accuracy the air resistance effect becomes important.

Take the advantage of repeating your experiment with a bouncy ball of your choice many times (in order to improve the statistical accuracy of your experiment as well as to take into account various false systematic effects).

Compare your results with tabular values for g in your area (note that values of the gravitational free fall acceleration vary as a function of local position).

Verify whether you can achieve with instruments like bouncy balls, an electronic chronometer on your smartphone and a standard ruler the following relative accuracy: 10^{-2}? 10^{-3}? 10^{-4}?

How could you further improve the accuracy (at least in principle)?

Thus, after performing this short treatise, which is inspired by Galileo Galilei, we were able to establish that the time t of free fall, or the period T of a bouncy ball, does not depend on the ball mass within the accuracy achievable in our simple experimental setups. Notice that we "measured" only integral characteristics of the ball motion, namely the initial height H of bounce and the bouncing period T, which are indeed very convenient experimental observables.

The fact established and verified throughout the present section is amazing and counterintuitive: as soon as all false effects are eliminated by means of cleverly designing your experiment, or at least they are properly taken into account and corrected for, you discover that various objects with different mass and composition fall down, or bounce on a surface, identically (that means here: an equal time for an equal fall height). However, let us be more ambitious.

Now we would like to explore the question of to what extent the conclusion on the universality of free fall is in fact universal. Is it a specific feature of motion near the surface of the Earth? Is it valid for objects falling down to the Sun? In other words, could we extend our finding to a more general case of motion in a gravitational field, say for the motion of planets and stars? Could we assume perhaps that the values of periods of planet orbiting do not change as a function of the planet masses?

Differential equations of motion of point-like masses ("*material points*"), which attract each other via the gravitational force, were discovered by

Isaac Newton[5]. They shed light on this question:

$$m_i \ddot{\vec{r}}_i = M_i G \sum_{i \neq j} M_j \frac{(\vec{r}_j - \vec{r}_i)}{|(\vec{r}_j - \vec{r}_i)|^3}. \tag{1.5}$$

In order to reveal the universal meaning of Galileo's discovery, we have to note two important messages, which are encrypted in Newton's equation.

The first message is that there are two distinguished properties of the mass. Namely, the mass m_i of i-th object that enters into the left-hand side of the famous equation (1.5), is the *inertial mass* (a measure of reaction of the object to an external action of other objects manifested in the change of its velocity), while the mass M_i on the right-hand side of this equation is the *gravitational mass* (a measure of the intensity of its gravitational interaction with other objects).

Note that the existence of two distinguished properties of the mass assumes the availability of two physically different measuring methods, which could not be reduced one to another. In fact, we could measure independently on one hand the gravitational mass of an object with a balance (static measurement, no motion is involved, the gravitational force is balanced by the spring resistance, for instance), and on the other hand the inertial mass (the acceleration of the mass is measured as a function of an applied force).

As stated in the Newton's Principia: "It [mass] can also be known from a body's weight, for – by making very accurate experiments with pendulums – I have found it to be proportional to the weight..."

The second message is that in fact these two distinguished properties of the mass enter into equation (1.5) in a degenerated manner: the value of inertial mass appears to be always equal to the value of gravitational mass: $(m_i = M_i)$. This latter statement is not trivial at all; such intriguing coincidences do not occur by chance, there should be a reason behind this regularity, and thus this equation is worth being reproduced here in the general form:

$$m = M. \tag{1.6}$$

---

[5]Newton, J. S. (1686). *Philosophiae Naturalis Principia Mathematica*, Imprimatur S.Perys, Reg.Soc. Praeses, Londini.

What is the deep reason for this coincidence?

An immediate formal consequence of these two observations is that an acceleration of an object, moving in the gravitational field of other objects, does not vary as a function of its mass $m = m_i = M_i$, because the inertial mass and the gravitational mass are precisely reduced on the left side and on the right side of equation (1.5) (provided that all *other-than-gravity interactions*, as well as its effect on other objects, could be neglected at the considered level of accuracy).

As the *"mass independence"* is valid for accelerations, would it be also valid for periods of motion?

In spite of the just-stated fact that the acceleration does not change as a function of mass, and also in spite of Galileo's statement on the universality of free fall, which we seemed to clearly confirm using bouncy balls, the answer is ... negative.

There is no way of extending accurately, in a general form, the conclusion on the mass independence of gravitational accelerations to the conclusion on the mass independence of *integral characteristics* of motion in a gravitational field, like the period of a planet's motion along its orbit, which is an analogue of a bouncing period. This is like that because the period of motion in the case of two interacting objects is a function of the *relative motion* of these two objects.

Consequently, it is easy to show using equation (1.5) that in the case of two gravitating masses their relative acceleration $\ddot{\vec{r}}_i - \ddot{\vec{r}}_j$ is proportional to the total mass $M_1 + M_2$ of the two-body system, thus the period is a function of the total mass as well (in the case of a larger number of masses a statement would become even more complicated). Only in the case of one mass being much larger than another mass, we would return to the statement on the independence of the period on the mass of the *light object*.

Problem 1.7. *Verify that the orbiting period of two gravitating masses does depend on their masses.*

Note: Use the gravitational law and the equation of motion.

Derive this dependence.

Explain why the orbiting period of a planet in the solar system can be approximately considered as independent of the planet's mass.

Find a correction to Kepler's law (the square of periods relate to each other as the cubes of the major semi-axis)[6] for the ratio of Jupiter and Saturn's periods due to accounting for the ratio of the planet masses to the Sun's mass.

Moreover, only as long as the mass of the bouncy ball is much smaller than the mass of the mirror (plus, of course, the total gravitating mass, which is rigidly attached to it) the period of the bouncy ball would change significantly as a function of the ball mass value. As gravity is extremely weak, and thus we need a huge mass in order to provide a significant effect of gravitation, the latter condition is met with an extremely high accuracy in all realistic laboratory experiments.

However, in a hypothetical case of comparable masses of a ball and a gravitating mirror, the bouncing period would depend explicitly on the mass values.

Problem 1.8. *Find a compact form of the equation of motion for a bouncy ball.*

Note: The following trick could help. Treat the motion of an object in the potential $Mg|z|$ from a height H to a height "below the surface of the mirror" $-H$.

Problem 1.9. *Verify that the ratio of squares of planet periods does change as a function of the planet masses.*

Find a correction to Kepler's law (the squares of periods relate to each other as the cubes of the major semi-axis) for the ratio of Jupiter and Saturn's periods due to accounting for the ratio of the planet masses to the Sun's mass.

Thus we could conclude that, in order to verify that the motion of objects in a gravitational field does not change as a function of their masses, we

---

[6]Kepler, J. (1619). *Harmonices Mundi, Johann Planck*, Linz, Austria, p. 189.

should measure rather the acceleration of these objects than the periods of their orbiting.

Are there any further complications and constraints in such measurements we have to take care of?

It is important to note that Newton's equation (1.5) is formulated for an acceleration of point-like massive particles, which interact with each other solely via the gravitational forces. However, as far as we introduce the finite size of physical objects or/and other-than-gravity inter-particle interactions, what would then be a consequence for the validity of the statement on the independence of an acceleration of an object in a gravitational field as a function of its mass?

In real gravitational experiments or observations we usually deal with spatially extended objects (like orbiting planets or mirrors and bouncing balls, for instance). Even in the case of measurements with elementary particles or atoms, for which their spatial extensions could be neglected with an extraordinary precision, another gravitating mass is always a classical macroscopic object. And the strength of the gravitational interaction might be different in different points of the object.

Then the acceleration of what point, say in a orbiting planet, should we measure?

Before answering this question, let us also take into account that real physical objects are made of parts (of atoms, molecules, larger constituents), which interact with each other also via forces of other-than-gravity nature (in particular via electromagnetic forces). The intensity of the electric force is proportional to electric charges, but not to masses. So our statement that the mass is excluded from the equation for the acceleration of a given part of the object is no longer valid.

The problem could be reformulated as follows: could we separate *internal degrees of freedom* (the relative motion of parts of the object) from the motion of its center of mass?

Would the motion of its center of mass be the same for different relative motions of internal parts of the object?

The general answer is negative!

We know very well some examples, in which such internal motions of the object's constituents do influence the motion of the center of mass of the whole object. One such example is the *tidal effect*, which slows down the rotation of the Earth due to the energy dissipation in the Earth's oceans during tides. (*Note*: Perhaps the reader would be surprised to learn that the same tidal effect accelerates the Moon and thus "pushes" it away from the Earth.)

Such a tidal effect is even better pronounced for those planets, which are located closer to the Sun than the Earth is, like for Venus and Mercury; it significantly influences their motion. To underline some extreme cases, we could mention that the tidal forces could tear out a star while it passes close to a black hole. And, of course, such effects are especially pronounced in the case of spatially extended objects (for instance, the galaxies interacting with each other).

Problem 1.10. *Verify that the motion of the center of mass of an object in an external gravitational field would not depend on the internal motion of the object's parts if the gravitational field were homogeneous within the object size.*

Thus we could conclude from the considerations given above, as well as from analyzing the above problem, that an object could be treated as a "material point" with no size, only as long as we are neglecting any non-homogeneities of a gravitational field within the size of the object, because in this case the motion of its center of mass would satisfy Newton's equation (1.5). Do such *homogeneous gravitational fields* exist?

The right-hand side of Newton's equation describes the gravitational field, which is produced by point-like sources. Unfortunately, the gradients of such a field are nonvanishing everywhere, and thus the field is never homogeneous. The only practical simplification of the problem in the typical experimental conditions follows from the extreme weakness of the gravitational interaction on one hand and the large size of nearest gravity sources on the other hand.

Therefore, a gravitational field could often be considered as homogeneous with high accuracy.

In order to clarify what we mean with the latter statement, let us consider the two following limiting cases.

If only one gravitating mass and a small trial object are present in the region of observations (i.e. all other masses are placed sufficiently far away from them so that their gravitational fields could be neglected at the considered level of accuracy), and also if the distance from the trial object to the center of the gravitating mass is much larger than the size of the trial object, then the gravitational field produced by the mass within the size of the trial object could be considered as quite homogeneous.

(*Note*: All corrections are typically not larger than at least the ratio of the size of the trial object to the distance from the center of the gravitating mass to the trial object.)

In contrast, if the size of the trial object is comparable to the distance to the center of the gravitating mass, then the gravitational field looks strongly inhomogeneous. It even changes its direction to the opposite one if the Earth is your gravitating mass, and the Earth's oceans are your trial object! (A simple manifestation of this effect is that the Earth is nearly round but it is not flat! If the field were homogeneous, the surface of the Earth would be flat. Does this picture remind you of something?)

Thus, in order to perform an accurate reliable test of the statement on the independence of a gravitational acceleration of an object as a function of its mass, we should take special care of such *gradients of a gravitational field*, and thus we should use only relatively small objects as probes and also we should place them relatively far away from the center of the gravitating mass. The following problems provide you with an idea of the degree of importance of these size effects.

Problem 1.11. *Assume that we have at our disposal an ideally flat horizontal disk surface with the diameter of 10 km; there are no vertical walls on the perimeter of this disk. Assume also that we could supply water into the center of this disk area as long as water is still confined on the top of it. If the gravitational field of the Earth were homogeneous then the surface of the circle would not be able to keep any significant amount of water in the static regime.*

How much water could be accumulated on the top of the flat horizontal disk surface due to the inhomogeneity of the gravitational field of the Earth?

Note: Look at figure [1.2]. It provides a kind of illustration for the present problem: the sea (or the lake) in the dreams of the cosmonaut, with the whale and the boat on its (truncated) spherical surface, are confined only by the gravitational gradient on the flat horizontal surface of the seashore with his wife and children waiting there for his return.

Note: Neglect here the effects of surface tension and explain why this is a valid approximation.

Estimate approximately the minimum diameter of such a truncated spherical lake, for which the effects of surface tension of water are still not very significant.

Problem 1.12. *Two cannonballs with the weight of each ball equal to 100 kg are placed respectively at the bow and at the stern of a 100 m long ship. The ship is sailing along the equator.*

Calculate the value of the force arising from the inhomogeneity of the gravitational field of the Earth, which "pushes" these cannonballs on the flat deck surface towards each other.

Problem 1.13. *A gradient of the Sun's gravitational field slightly "polarizes" the Moon's orbit.*

At what mean distance from the Earth could any its satellite be torn out from the Earth by the Sun?

These several examples illustrate that our ability to formulate a statement on the independence of the motion of physical objects in a gravitational field as a function of their mass, which would be universal on one hand and broad on the other hand, is rather compromised. For instance, as soon as we take into account the effects of gravitational gradients or non-gravitational degrees of freedom inside the objects, like the tidal effect, this statement is violated.

In other words, as soon as any realistic physical object is treated, the mentioned statement is always violated, strictly speaking.

Is there a statement which would still remain valid after this critical analysis?

The most accurate and universal statement that we know is a rather abstract statement on the equality of inertial and gravitational masses of point-like objects ("material points"), which is given in equation (1.6). The motion of the inertial mass is defined in an *accelerated frame of reference*, while the motion of the gravitational mass is defined in a gravitational field. Strictly speaking, their geometries are the same only for homogeneous gravitational fields, which does not exist...

The equality of inertial and gravitational masses of point-like objects is the content of the famous equivalence principle[7], which is, in a turn, the cornerstone of the general theory of relativity.

Such a universal statement as the equivalence principle should be formulated for *intrinsic properties* of particles (in classical physics a particle means "a very small object" such that its size could be neglected), such as the inertial mass m (a measure of reaction of the object to an external action of other objects manifested in the change of its velocity) and the gravitational mass M (a measure of the intensity of its gravitational interaction with other objects).

The equivalence principle postulates the equality of these two masses. The independence of the acceleration of an object as a function of its mass is a *consequence* of this principle, but it is not the principle itself.

A high accuracy test of certain consequences of the equivalence principle for any realistic object is possible only via observations of the motion of such an object; the object size, the inhomogeneity of the gravitational field within the object size, all internal degrees of freedom, the interaction with other objects, and so on, cannot be easily excluded from such an experiment; however, they have to be suppressed and the residual effects have to be accounted for in order to avoid undesired false effects.

Moreover, at the ultimate level of accuracy that we could in principle achieve, interactions of a different nature start to be interrelated. A future theory, which aims at unifying all kinds of fundamental interactions, should address the question whether the equivalence principle would be still valid (or violated) at that level, or probably replaced by a more general principle.

Other questions arise as soon as we are going to extend the equivalence principle to the world of microscopic objects, where, as we have mentioned in the preface, wave properties would clearly manifest themselves, and the "material point" concept could not be applied anymore.

What are the observable quantities that we could use for experimental tests of the equivalence principle in the microscopic world?

How would the wave nature of quantum microscopic objects correct our classical results?

Before discussing all these questions we invite the reader into a falling elevator.

---

[7]Einstein, A. (1908). Berichtigungen zu der Arbeit: ber das Relativittsprinzip und die aus demselben gezogenen Folgerungen, Jahr. Rad. Elektr. **5**, p. 98.

1.2   Sailing Ship, Flying Jet and Falling Elevator: The Principle of Relativity

"Concepts that have proven useful in ordering things easily achieve such authority over us that we forget their earthly origin and accept them as unalterable givens."[8]

Albert Einstein

Since ancient times, motion, in contrast to rest, was considered as an important means for recovery. This statement sounds even more solemn in Latin: "Mobilitate viget viresque acquirit eundo" means something like "It grows by moving and gathers strength as it spreads on". It fixes clearly the difference between rest, which is a natural state and thus it does not require any effort, and motion, which always requires a lot of effort but rewards us with physical and mental health.

However, following the same logic as in the case of *free fall*, Galileo Galilei concluded that the difference between *free motion* and *rest* is also an illusion.

This conclusion allowed him to formulate a crucial principle, which later became known as the *principle of relativity*[9]. It assumes that as long as you are in a closed room, for instance in a cabin on a ship that is moving uniformly, there is no experiment in framework of classical mechanics that would allow you to distinguish whether the ship is actually moving: such mechanical phenomena should look the same in all *inertial frames of reference* (i.e. in reference frames which move uniformly).

Imagine you are a happy participant of the scene in figure [1.1]. Close the door, take the ball and drop it to the floor. The bouncing would not be affected by the fact that the door is closed. Imagine now that the whole room is smoothly and uniformly moving, being for instance on board of a ship. You cannot recognize the presence of this motion because you could not see the Pisa tower through the open door anymore. But you also could not recognize the presence of this motion, because the bouncy ball would also not be affected by the (uniform) motion of the room.

Later this principle of relativity was generalized by Albert Einstein to

---

[8]Einstein, A. (1916). Ernst Mach, Physikalische Zeitschrift. **17**, p. 101.

[9]Galileo (1638). *Discorsi e Dimostrazioni Matematiche, intorno a due nuove scienze*, Lowys Elzevir, Leiden, p. 191.

all physical laws[10].

Isn't it surprising that one could make a statement about any physical law, even about one as yet unknown?

This formulation of the principle of relativity suggests an important general procedure. Namely it assumes that (all) physical phenomena should be described independently in different frames of reference firstly and then the descriptions of (each) the same physical phenomenon observed in different frames of reference should be compared to each other. This is a fruitful and deep approach with far-reaching consequences, which we have to discuss in more detail.

For the moment we are going to leave the times of Galileo Galilei, when ships were widely used, for more recent times of elevators. Some people, who have happened to be in a falling elevator, tell about their exciting experience of weightlessness. You probably know that one could experience the same unforgettable feeling (in safer conditions) in Disneyland or in a special "falling plane", where astronauts prepare themselves for the weightlessness in orbit flights.

Let us imagine that we have a chance to be inside an elevator, which is falling down with the acceleration that is a bit smaller than the standard gravitational acceleration (this condition is rather natural for a freely falling object, because it is normally met due to the presence of some friction or air resistance, which slows the object down a bit). Also, we assume that the elevator is going to continue falling down for a sufficiently long time, so we could play with a bouncy ball.

What would happen with the ball at different phases of the elevator's fall? How the ball, which is bouncing on the floor of the elevator, would be affected by the weightlessness?

We drop the ball on the floor of the elevator and observe that the period of its bouncing is much longer than it is usually (this is not so because every second in a falling elevator seems like an hour for you!). Perhaps there is also a balance in the elevator, and a weighting shows you that both the bouncy ball and the experimentalist lose a major fraction of their weight. We remember that both the period and the weight are functions of the value of the gravitational strength g.

Then there is no other conclusion we could draw from our observations except for the statement that the gravity strength decreased significantly

[10]Einstein, A., Lorentz, H. A., Minkowski, H., and Weyl, H. (1952). *The Principle of Relativity: A Collection of Original Memoirs on the Special and General Theory of Relativity*, Dover Publ., Mineola, p. 111.

in our elevator.

Suddenly something happened. The gravity strength started to increase, then it rapidly reached its normal value and promptly became even larger than the standard value. As a manifestation of this change, the bouncing period decreased, while the balance showed an increase in the weight. The reader has already guessed that at this moment the elevator's fall ended, but inside the elevator we would feel only a fast increase of the gravity strength.

Exercise 1.8. *Perform the experiment described above (drop a bouncy ball to the floor of an elevator) in a normal – not falling! – elevator, and observe the behavior of this bouncy ball at different phases of the elevator's motion.*

Using the bouncy ball and a clock, try to quantify approximately the values of acceleration of the elevator at different phases of its motion.

Measure the values of acceleration of the elevator at different phases of its motion using a balance.

Compare these two results (those obtained with the bouncy ball and with the balance).

The moral of this story is that as long as the doors of the elevator are closed and the walls, the doors, the floor and the ceiling are not transparent for any kind of external signals, it is not easy to distinguish a local gravity and an acceleration of the elevator. (*Note*: It is not easy at all to "design" even an imaginary elevator, in which you would be completely isolated from *all* external signals. Think of the electromagnetic radiation of a broad frequency range, high-energy elementary particles, neutrinos, and so on.)

The equality of gravity in a given point and the inertial force in an accelerated frame of reference for *any* particle is sometimes used as another formulation of the equivalence principle.

(*Note*: However, one should be careful with this formulation; we are going to show below that this formulation could be rather misleading. The problem follows from the fact that physically important properties of a gravitational field are not only those in a given spatial point, but also those in its close vicinity. As far as the "geometry" of a gravitational force differs from the "geometry" of an inertial force, in principle they can always be distinguished.)

A comparison between the two above mentioned examples, a uniformly moving ship on one hand and a freely falling elevator on the other hand, demonstrates not only their important *differences* (uniformly moving and accelerated frames of reference, respectively) but also their *similarity*. Indeed, both examples deal with physical laboratories, in which experiments of any kind can hardly distinguish certain types of motion.

Let us look at the problem of uniformly moving frames of reference more thoroughly. In order to enhance a phenomenon, which is going to be of interest for us here, we would like to consider a much larger velocity. Thus we would like to benefit from the achievements of modern civilization and place our laboratory inside a supersonic jet that is uniformly flying. There is no better way to spend your time on a transatlantic flight than to drop a bouncy ball from a fixed height and to measure its bouncing period carefully.

After conducting a few tests in jets of different airlines (usually they all fly at a similar height of about 10 km above the ground level at a similar ground speed of some 800 km/hour, so the conditions of your different measurements are close to each other) you would probably conclude that the bouncing periods are equal to each other within the level of accuracy that you could achieve in your simple measurements. The result should not change as a function of the jet velocity and the flight direction.

In such simple test experiments it would be difficult to avoid numerous systematic false effects. However, under certain conditions, with a very good elastic bouncy ball and an accurate enough study you would discover that on top of the well-known gravitational force there is another tiny force, which acts on the bouncy ball and changes as a function of the jet's velocity value and direction.

This tiny force is known as the *Coriolis force*[11]; it was discovered much later than Galileo's treatise of the principle of relativity (one reason for this delay is probably due to the fact that several potentially observable effects caused by the Coriolis force associated with the rotation of the Earth, were relatively small in the Galileo's time as the force is proportional to the velocity and velocities were lower). This force, $\vec{F}_c$, has the following mathematical form:

$$\vec{F}_c = -2m[\vec{\omega} \times \vec{v}]. \tag{1.7}$$

---

[11]Coriolis, G. (1835). Sur les équations du mouvement relative des systmes de corps, Journ. Ecole Polytechn. **15**, p. 142.

Here m is the mass of the bouncing ball, $\vec{v}$ is the velocity of the ball measured *relative to the ground* (so the velocity of the jet is a component of the total velocity; and evidently the jet velocity is its dominant contribution). (*Note*: The symbol × is used in order to indicate the cross product of the two vectors $\vec{\omega}$ and $\vec{v}$. For the readers who are not acquainted with the concept of cross product ×, we would like to mention that the result of a cross product of two vectors is a vector, which is normal to both these vectors $\vec{\omega}$ and $\vec{v}$, with the absolute value given in a factor $wv\sin(\Theta)$, with Θ being the angle between the vectors $\vec{\omega}$ and $\vec{v}$.)

But what is the physical sense of this mysterious vector $\vec{\omega}$ in the above expression? After performing analogous experiments in different places at different velocities of motion of the frame of reference, we would be able to find out that it is a vector, which is always directed parallel to the Earth's rotation axis from the south pole to the north pole; with the same value in any point on the surface of the Earth, which is equal to the angular frequency of the rotation of the Earth.

Problem 1.14. *An elastic ball is bouncing with some frequency on the floor of a supersonic jet, which stays landed in an airport. Then the jet departs and someone drops the ball to the floor again.*

Estimate the change in the bouncing period (arising due to the Coriolis effect) provided that the jet is flying at the ground speed of 3000 km/hour, in two cases:

i) The jet is flying along the equator from east to west.

ii) It is flying from west to east.

Exercise 1.9. *Estimate an accuracy of the bouncy ball experiment in the previous problem with a supersonic jet, which is required for clearly detecting the difference between the two mentioned cases.*

Would it be feasible to achieve such an accuracy with a good elastic bouncy ball?

Justify your conclusion.

Exercise 1.10. *Repeat analogous calculations for an experiment per-formed by an observer on board of a submarine moving at the ground velocity of 30 km/hour.*

What is your conclusion concerning the feasibility of observing the Cori-olis force using a bouncy ball on board of the submarine?

Does the inevitable existence of the Coriolis effect mean that Galileo's prin-ciple of relativity is violated?

The affirmative conclusion might seem to be rather natural. However, it is not that straightforward. In fact it is wrong. The right answer is negative. To be formal, the accurate meaning of Galileo's principle of relativity is that there exists such a place in the Universe, where uniformly moving frames of reference are equivalent to each other (they can not be distinguished in physical experiments). These are inertial frames of reference, while our Earth is not.

Why is this so? The Earth rotates around its own axis. This rotation is characterized by the vector $\vec{\omega}$, which is present in any spatial point in any frame of reference connected rigidly to the rotating Earth, and this vector plays the role of an effective field, which affects all moving objects. Moreover, the Earth rotates around the Sun; simultaneously, the Sun ro-tates, together with the whole solar system including the Earth, around the center of our galaxy, and so on.

In other words, the principle of relativity states that there are no such fields, which pierce through empty space, when we are found in those spe-cial inertial systems. An attentive reader would notice that even very far from the Earth something is always rotating and thus producing effective forces. Also such phenomena, as relict electromagnetic (and also probably gravitational) radiation is always there, thus providing a kind of absolute reference frame.

However, we could treat them as "background", choose a nonrotating system and not include the relict radiation in our definition of the reference frame. In fact, this statement is a "hidden" assertion about the *properties of vacuum*! (*Note:* The term "vacuum" usually denotes the space with no substance in it. The meaning of vacuum in quantum theory is more complicated than that. Here vacuum is the lowest possible energy state of fields, which fill our world.)

This statement brings into consideration *free empty space* (space that we observe while describing phenomena in inertial frames of reference), in which many physical laws start looking simpler, and various so-called *effective forces* related to the choice of a frame of reference do not show up. Isn't it curious that the principle of relativity in this interpretation introduces a kind of absolute reference in the Universe, an inertial frame of reference?

Let us investigate in more detail how the principle of relativity establishes the relation between different inertial frames of reference and how the difference between inertial and noninertial frames of reference manifests itself.

For that purpose let us return to the ships in the sea and an experimentalist on the shore. Suppose that some interesting event happened on the shore, which we would call a reference frame K, in a point with coordinates $\vec{R}$ at a time instant t. For example, a bouncy ball was dropped. If we look from the ship, which is uniformly moving at a velocity $\vec{v}$, with the associated reference frame denoted K', the same event would have coordinates $\vec{R}'$, while a time moment is t'.

Could we predict what the bouncing on the shore (K) would look like from the ship (K')?

We have learnt already the law of bouncing of a ball on the shore. Now we would like to find a *transformation rule* between the reference frame R, t and the reference frame R', t'. The actual relation between coordinates $\vec{R}, t$ and $\vec{R}', t'$ in these frames of reference *has to be a matter of experimental finding*. Although our experience suggests to us that a grand lady on a shore does not look to an observer in a moving ship like a young girl, this guess could also become an illusion.

In fact, it is!

The special theory of relativity (which we are going to mention again very soon) states that only as far as a ship's velocity is small compared to that of light we could accept intuitively the truly geometrical rules, which are known as the *Galileo transformations*:

$$\vec{R}' = \vec{R} + \vec{v}t, \tag{1.8}$$
$$t' = t. \tag{1.9}$$

It follows immediately from the principle of relativity that the form of the equation of motion of a bouncing ball should not change while coordinates and time in a K-frame of reference are substituted by coordinates and time in a K'-frame of reference using the transformation rules (1.8, 1.9) (and an analogous statement should be also valid for all other physical laws). In order words, it is said that all physical laws should be *covariant* under the Galileo transformations.

Newton's equation of motion is covariant under the Galileo transformations because the Galileo transformations include only *relative distances* (between interacting masses) and time intervals (between events). However, they do not include their absolute values. At the same time neither relative distances between objects nor time intervals (and thus also neither relative velocities nor relative accelerations) change under such transformations; such values are called *invariants*.

Consequently, the bouncing ball period changes as a function of the drop height also in accordance with a covariant law (neglecting in this statement the relativistic effects). If Newton's equation included noncovariant forces, such as the Coriolis force, it would not be covariant under the Galileo transformations. That is why these equations are valid only in inertial frames of reference, in which no Coriolis force or analogous force exists.

This observation means that the principle of relativity, supplemented with the rules of transformations between coordinates and time in inertial frames of reference, plays the role of a referee for checking the validity of any new hypotheses: only such a hypothetical law could be valid, the mathematical formulation of which is covariant under the transformation rules; otherwise we can state that a new hypothetical law is wrong even without going into further details.

In fact, a standard approach in modern theoretical physics consists of searching for a new physical law by means of constructing some corresponding equations from covariant values, very much like a building is constructed using prefabricated blocks. The principle of relativity could also be used vice versa: once a physical law is established experimentally, then the transformation rules, which would keep this law covariant, could be derived mathematically.

In particular it was noticed that the laws of electrodynamics, while being well established experimentally, are surprisingly not covariant under the Galileo transformations. Indeed, a charged particle might experience the action of either pure magnetic or also electric fields as a function of the choice of an inertial frame of reference. This strange asymmetry forced physicists to look for the transformation rules, under which the laws of electrodynamics are covariant. These rules are so-called *Lorentz transformation rules*[12].

What was a result of this finding? The rejection of the principle of relativity? No, something else. These were the Galileo transformation rules which are substituted by the Lorentz transformation rules.

The relation between space and time in different inertial frames of reference, given in the Lorentz transformations, turned out to become later the main content of the Einstein's special theory of relativity. This relation between coordinates X, t in one frame of reference and coordinates X', t' in another frame of reference, which is moving relative to the first one along X-axis at a velocity v, includes the speed of light c as its important ingredient.

The Lorentz transformation rule looks like:

$$X' = \frac{X - vt}{\sqrt{1 - v^2/c^2}}, \tag{1.10}$$

$$t' = \frac{t - vX/c^2}{\sqrt{1 - v^2/c^2}}. \tag{1.11}$$

Problem 1.15. *A ball is bouncing with a period T on the ground. A camera in a spy satellite, which is moving at the ground speed of 8 km/s, is observing this ball from the orbit.*

What value of the period of the bouncing ball would be recorded by the camera?

Exercise 1.11. *Replace the satellite in the above problem with a ship floating in the sea at the velocity of 30 km/hour.*

What value of the period of the bouncing ball would be measured by an observer in the ship?

We are not going to discuss here a rich world of all consequences of the above relations because this discussion would deviate too much from the

---

[12]Lorentz, H. A. (1899). Simplified theory of electrical and optical phenomena in moving systems, Proc. Royal Netherlands Acad. Arts Scie. **1**, p. 427.

main subject of our book. However, it is worth mentioning that the idea that the time rate and the spatial scale change as a function of velocity, as stated in the Lorentz transformation rules, was an absolutely revolutionary concept, while a combination of the Lorentz transformations with the principle of relativity became one of the most fruitful ideas in the physics of the twentieth century.

But what is wrong with accelerated frames of reference?

Why are the accelerated frames of reference deprived of equal rights with inertial frames of reference?

A clear difference between these two cases, between inertial and accelerated frames of reference, is established in Newton's laws. Newton's laws state that the reason for acceleration of an object in an inertial frame of reference could only consist of its interaction with another object. To turn this formulation differently, it is said that the motion of a free object in an inertial frame of reference is the motion along a straight line at a constant velocity.

Important questions arise: what is *a straight line* for a physical experiment and what is *a uniformly running clock*?

A physical definition, which is consistent with Newton's laws, is the following: the straight line is defined as a line, along which light propagates in empty space, and the clock is a device built of a pair of parallel mirrors, between which a light pulse oscillates. The coordinate system is imagined to be realized using absolutely rigid rods, which could be (at least in principle) however long. So an experimental test of the inertial character of a frame of reference could consist of observing the motion of a free object.

Problem 1.16. *Describe a trajectory of an object, which is sliding without any friction on a surface of a flat horizontal floor in an arbitrary initial direction.*

Note: Take into account the existence of the Coriolis force and neglect the air resistance.

Note: The flat horizontal floor is located in your present place for concreteness.

Problem 1.17. *Describe a trajectory of motion of an ideal bouncy ball ("ideal" means here: no energy dissipation, no friction, no rotation, no* *initial velocity, etc.), which is dropped from the height H to an ideal flat horizontal floor.*

Note: Take into account the existence of the Coriolis force.

Note: The bouncy ball experiment is located in your present place for concreteness.

Problem 1.18. *Analyze whether a horizontal drift of the bouncing ball in the above problem could be compensated by means of inclining the flat floor by some angle.*

If not, explain why.

If yes, calculate the angle of the floor inclination needed.

Note: This experiment is also located in your present place for concreteness.

Problem 1.19. *Replace the horizontal flat floor in the previous problems with a spherical surface with the center of this sphere, which coincides with the center of the Earth, and describe a trajectory of the bouncing ball in such a system.*

Compare solutions of the three above problems for these two cases: with a flat floor and with a spherical floor.

(*Note:* A method to establish that the trajectory is a straight line consists of looking along this line to a distant [i.e. point-like] source of light. A more elaborated practical approach consists of using a laser pointer. However, the accuracy of this method would be compromised because of the angular divergence of light rays, which could not be kept however small. We are going to see below that this limitation is of principal character, which follows from the wave nature of light and all objects around.)

If you have good reason to think that a trial object is located far enough from all other gravitating masses that could noticeably affect its motion, i.e. if the trial object is considered to be free, and if, at the same time, you measure that it is moving straight along the line of the light propagation, then you might assume that the frame of reference is an inertial one. After everything stated above, this conclusion seems reasonable. However, is only this conclusion possible?

If the trial object in the preceding example is moving along a curved trajectory (relative to the trajectory of the light propagation), what could you conclude then? Could you state that the frame of reference is not an inertial one and thus some extra effective forces appear in this frame? Or perhaps the trajectory is curved because of the action of the gravitational force generated by another massive object, which we cannot see directly (however it is there)?

In order to analyze these questions in more detail and to show that there are more logical options in these considerations, we would like to return to a falling elevator. Imagine that, while being in the limited space of your closed laboratory (elevator), you are not aware about its actual motion. However you have observed in various experiments that any trajectory of a free object is always a straight line. Then what would you conclude about the presence of gravity?

Of course, in order to fool you around, someone super-powerful could purposely place this elevator in a place free of gravity somewhere far from the Earth without letting you know about the joke. However, you could also reasonably assume that the Earth's gravity is there, but it is locally eliminated in a frame of reference rigidly connected to your freely falling elevator (below we are going to specify this statement and comment on the domain of its applicability).

From this point, there are only two steps to the main postulate of the general theory of relativity.

A major step consists of hypothesizing that a gravitational field in *a given point* is equivalent to a transition to a locally accelerated frame of reference. The deep sense of this statement is that gravity can be reproduced (imitated) by means of a *transformation of space and time* (like this is done in the case of a transition from one frame of reference to another frame) and thus gravity could be imagined as a certain property of space-time. This statement is a direct consequence of the equality $m = M$.

The final step consists of converting the previous statement into a mathematically strict formulation, namely it consists of establishing that in fact gravity is described in terms of space-time transformations of a more general type than the transformation to a locally accelerated frame of reference. Gravity is then related to a curvature of space-time, i.e. to a purely geometrical property of space-time which determines in particular a form of the *shortest path* between two points.

The latter concept assumes that the motion of any object could be described in such new curved space-time coordinates in a way analogous to the free motion of an object in an inertial frame of reference; namely, it could be represented as a motion along the shortest path between two given points. And the shortest path between two points is defined with a curvature of space-time, which is the same for any trial object (neglecting an extra curvature induced by the trail object itself).

However, in curved space-time, which would fully reproduce all observable effects of gravity, these shortest paths are no longer straight lines, as well as the uniform clocks are no longer such. In particular, if we measure a sum of internal angles in a triangle, defined by three light rays, this sum would differ from π (radians) in such a curved space (experimental observations confirmed that the light rays are deflected by massive stars, so that the sum of three internal angles in a corresponding light triangle is no longer equal to π in presence of massive objects).

Thus one could say that gravity is an illusion which has risen due to the *geometry of space-time*!

One has to comment on the role of the statement $m = M$ and on the sense of the word "*local*" in the formulation of the equivalence principle understood as a means to locally eliminate gravity in a freely falling frame of reference. The equality of the gravitational mass M and the inertial mass m for any physical object allows the substitution of the terminology of forces, acting on a given object, by the terminology of space-time transformations, acting universally on all objects.

Thus, the equality of gravitational and inertial masses $M = m$ is an absolutely necessary condition for the application of the above mentioned geometrical interpretation of gravity. In the case of *a special type* of space-time transformations, namely those equivalent to the transition from an inertial frame of reference to a frame of reference rigidly connected to a freely falling elevator, the condition $M = m$ allows us to eliminate gravity at a given point.

However, the nature of a gravitational field cannot be exhausted by this equality. There are also other important properties of a gravitational field, which are described by its behavior in a nonvanishing volume of space around.

In particular, Newton's law of gravity establishes that the gravitational force is directed towards the source of gravity, and thus, due to the additivity of contributions from different sources, it is directed to the center of mass. Thus, if we consider, for instance, the gravitational attraction of a small trial object to a spherically shaped uniform massive body, the force lines are directed towards the center of the body. This property of the field is described by a certain behavior of its *gradients*.

Thus a curvature of space-time is required in the geometrical theory of gravity in order to describe properly the field properties in nonvanishing space volumes.

In terms of space-time geometry, the statement on the elimination of a gravitational field in a given spatial point related to a freely falling frame of reference by means of an acceleration of this freely falling frame of reference is equivalent to the simple fact that properties of a continuous curved surface in the small vicinity of the given point on the surface are mathematically equivalent to those of a plane which is tangential to this surface in the given point.

However, it is not sufficient to know locally the properties of a plane in order to describe the whole curved space globally. In physical terms this statement means that a gravitational field is defined not only by its value in a given spatial point but also by its gradient in this spatial point. Thus gravity could be distinguished from inertia in a freely falling frame of reference as soon as we are capable to measure gradients of the gravitational field.

It is interesting to find out what we could learn about gradients of a gravitational field while staying in a restricted volume of an elevator isolated from the external world.

Gravity is not shadowed by the walls of the elevator, it penetrates through them. Therefore we could try to perform accurate measurements of the gravitational forces, which act in *different locations* inside the elevator. For instance, we could investigate the motion of small pellets or water drops, which would fall down along the gravitational force lines. We remember that a gravitational force, in contrast to an inertial force, is characterized by its unavoidable gradients; so we could try to detect them.

However, the smaller the elevator, the more accurately such a measurement should be performed in order to reveal the gravitational field gradient. Unfortunately, such a simple experiment with freely falling water droplets would be hopelessly inaccurate. In order to improve the accuracy, could we profit somehow from the major increase in sensitivity, which we have already discovered from experiments with bouncing balls? And to what extent?

In order to address these questions, let us first look to figure [1.2]. As the reader knows, a nearly spherical shape of the Earth is due to the gravitational force, which attracts any object on its surface towards its center.

Fig. 1.2 Illusion of gravity.

And thus the Earth's oceans, with all ships on their surface and all fishes in their depths, are also of a spherical shape independently of the shape of the ocean bed as it is clearly reflected in the dreams of the cosmonaut in figure [1.2].

The women and the children, who are probably waiting on the Earth for the cosmonaut to return, are standing on the flat horizontal ocean shore. If the ocean bed is also flat and horizontal and constitutes just a continuation of the shore, the ocean surface would be approximately spherical. If the

ocean bed is concave or convex or of any complex shape, the result would
be about the same because the major part of the gravitational force is due
to the huge bulk of the rest of the Earth, so local details are of minor
importance.

As the reader has probably noticed, water forms a *spherical cap* on the
top of a flat horizontal surface of the table in the cockpit of the spaceship.
However, the proper interpretation of nature of this aquarium depends on
actual external conditions. Is the spaceship still standing on the surface
of the Earth? Is it flying somewhere in space far from large gravitating
masses? What is its acceleration in terms of the absolute value and direction
relative to the residual gravitational field?

If the spaceship is still on the ground of the Earth, then the aquarium
surface formed under the effect of gravity of the Earth is shown in the figure
in an exaggerated artistic manner. The shape of the water's surface on the
table top has a direct relation to the spherical shape of the Earth's oceans.
The radius of curvature of these two surfaces would be precisely the same
if the effects of surface tension would be absent (or small enough). And
there would be no space for fishes and sea horses in such an aquarium.

Nevertheless, what is the nature of the force, which pushes water towards the center of the horizontal table, and which pushes water towards
the center of the flat horizontal ocean bed?

It is the gradient of the gravitational field!

As the gravitational force lines are always directed towards the center
of the Earth, there are two components: one component is perpendicular
to the table's surface, while another one is directed towards the center of
the table at any point on the table. As is clear from simple geometrical
arguments, the strength of the latter component is smaller than the gravitational acceleration "g" by a factor, which is about equal to the ratio of
the radius of the Earth to the distance from the center of the table.

For a typical table size, this ratio factor is equal to about 10 million,
thus the height of the cap would be negligible and the effect would be
indistinguishable on top of the effects of surface tension. However, if the
whole laboratory is placed inside a falling elevator, or inside a freely falling
spaceship, everything would be different. The perpendicular component of
the gravitational acceleration would be compensated by an acceleration,
while the horizonal component would not.

In the case of a major fraction of gravity being compensated by an
acceleration, a measurement of the height of the water cap using a usual
ruler would give you a value of the curvature of the residual gravitational

field. If you replace water with a bouncing ball, the result would be similar in some sense: this cap would correspond to the line of trajectories traced by the bouncing ball, as it follows simply from the conservation of energy (assuming that the components of motion of the ball can intermix).

Therefore, the cosmonaut drawn in figure [1.2] would have a convenient instrument to measure gradients of the gravitational field even if this gravitational field is produced by some invisible dangerous gravitating mass. And the cosmonaut would be able to trace safely his path between such attracting masses in order to escape the danger. A useful device for cosmic explorations, which prevents you from eventually falling down into a black hole!

Let us return to the elevator at rest and to a bouncy ball at your disposal. Could we observe some residual small gravitational gradients even in this case?

The gradient force is small, but it is not negligible. While an effect of the gradient on the period of vertical bouncing would be impossible to observe with a simple bouncy ball, the gradient of the horizontal force might cause a potentially measurable effect as we are going to conclude after solving a few problems given below. An important experimental advantage follows from the fact that there are no major *background* forces present in the horizonal direction in a properly designed experiment.

One should not underestimate, however, all experimental complications associated with such an experiment (the degree of horizontality and flatness of the floor needed, an incredible control of any systematic effects related to the facts that the bouncy ball and the floor are not idealized but real physical objects with related asymmetries and irregularities, that rotations of the ball could not be neglected at this level of accuracy, that the Coriolis force is there, and so on).

Nevertheless, analogous measurements with quantum particles described in the following chapters could be more realistic, more statistically sensitive and free of major systematical effects inherent for mechanical elastic bouncing balls. Therefore, let us estimate the characteristic parameters of the problem by means of solving the following few problems. These characteristic parameters are going to be the same for bouncing balls and for bouncing elementary particles.

Problem 1.20. *In the problem described above, the gravitational gradient force pushes any object on the top of an ideal horizontal flat table towards the center of the table.*

Show that the gradient force could be described with a high accuracy in the model of a harmonic oscillator.

Estimate the period of horizontal oscillations of an ideal bouncy ball on top of the table.

Note: Neglect here the Coriolis force, and imagine that the table is placed on a pole.

Problem 1.21. *Describe the shape and the angular orientation of the mirror, which should be used in the above problem in order to cancel the gradient force.*

Note: Neglect the Coriolis force.

The method of measuring gradients of a gravitational field, which could provide us with a powerful tool to always distinguish inertia and gravity, is an example of a *global observation*. Although this statement might seem counterintuitive, a measurement of a gravitational field gradient in a restricted (but nevertheless finite) volume of a closed elevator is a global observation in contrast to the idealization of a hypothetical measurement on a point.

Any actual measurement is performed using an experimental setup of a finite size; thus the gravitational field is not uniform within its size. Therefore the gravitational gradient could be measured in any actual experiment, at least in principle, while locally (which means that we would be able to measure a value of the force in a given point) a gravitation force in an inertial frame of reference cannot be distinguished from acceleration of a frame of reference.

Nevertheless, one should always remember that as the gravitational field in typical conditions of laboratory experiments is weak and the characteristic spatial scale of gravitational gradients is large, the gradient effects are relatively weak.

In the case of the ceiling or walls of the elevator being made of glass, we could observe the motion of distant objects like stars, which all move with an equal acceleration. It would contradict the law of gravity, which says that a gravitational force decreases as the square of distance, so it is unlikely that all objects move with an equal acceleration. This is an indirect method to distinguish an inertial frame of reference and an accelerated one using *global* observations.

If we observe that faraway stars move with an equal acceleration, if we detect that a free object moves with an acceleration (in particular it moves along a curved line), but we could not point out another object (even located far away) which causes this acceleration or curvature, then we conclude that our frame of reference is not an inertial one. In the opposite case we could say (at least with a given accuracy of our global observations) that our frame of reference is an inertial frame.

Problem 1.22. *The gravitational attraction of the Moon to the Sun is stronger than the gravitational attraction of the Moon to the Earth by approximately a factor of 2.7.*

Why is the Moon still there?

Problem 1.23. *You are standing on the top of a sole cubic rock with a size of 600 m, which is located in the middle of an ocean; you are holding a long rope with a weight on its end such that the rope serves as a plumb-line (you could imagine that you are going to catch a fish from the top of the rock).*

What would be the gravitational deviation from the vertical direction (understood as the direction towards the center of the Earth, or as the direction of free fall provided the absence of the rock) of the weight on the end of the 600 m long rope?

Note: Assume for simplicity that the Earth and the rock mean densities are equal to each other, also that the rectangular cross section of the rock continues deep under the water's surface.

Problem 1.24. *You drop a small ball from the top of the rock, described in the previous problem. The ball is initially positioned at a very small distance h from the flat vertical wall of the rock.*

What is the trajectory of the falling ball?

Under what conditions could one observe bouncing of the ball from the vertical wall of the rock?

Note: Ignore for simplicity the air resistance and the existence of the Coriolis effect.

Let us summarize our findings.

We understood that inertial frames of reference are distinguished among other frames due to the fact that the reason for an accelerated motion of a trial object described in an inertial frame could consist only of its interaction with other objects, which we could in principle point out.

If we do not see any mass that could deflect the motion of our trial object from a straight line, and if at the same time the trajectories of *all* trial objects diverge from the uniform motion along a straight line, then we rather assume that there is a gravitational field of an invisible object, or that our frame of reference is accelerated.

The universality of the effects of gravity on all point-like objects ("material points") makes them indistinguishable from the effects of inertia in an accelerated frame of reference measured at a given point (*locally*).

The imminent presence of gradients of a gravitational field allows us to distinguish gravity and acceleration of frames of reference due to *global observations*.

Properties of the *global* gravity field could be described within the concept of curved space-time.

In the next section we are going to study our illusions concerning the flatness of our space and the motionless of our frame of reference, and also we are going to see how a curvature of space generates gravity.

1.3 Artificial Gravity and Curved Space: The Least Action Principle

"When I was in high school, my physics teacher – whose name was Mr. Bader – called me down one day after physics class and said: "You look bored. I want to tell you something interesting." Then he told me something which I found absolutely fascinating, and have, since that time, always

found fascinating. The subject is this – the principle of least action."[13]

Richard Feynman

Imagine that you are waking up suddenly in the darkness of the night inside a large *centrifuge...*

In fact, we even do not need to imagine something like that, as we are indeed always there!

Exercise 1.12. *Estimate the absolute value of the centrifugal accelera-tion, which acts on an object standing on the surface of the Earth; take into account only the acceleration resulting from the rotation of the Earth around its own axis.*

Compare the absolute values of centrifugal and gravitational accelera-tions.

Calculate the angle between vectors of gravitational and centrifugal ac-celeration.

Note: For concreteness, assume that the object is placed in your present location.

Exercise 1.13. *Imagine that the period of rotation of the Earth in the above exercise would be larger than it really is. Imagine also for a moment that the Earth would be so rigid that its shape stays unchanged even at larger rotation frequencies.*

Estimate then the duration of the Earth's day, which corresponds to so high a frequency of rotation of the Earth that the centrifugal force would become as large as to totally compensate the gravitational force, or in other words an object on the surface of the Earth would be found in a state of weightiness.

Although this centrifuge is really there, nevertheless we do not think about it. We do not feel ourselves living inside this huge centrifuge at

[13]Feynman, R.P., Leighton, R.B., and Sands, M. (1964). *The Feynman Lectures in Physics, v. 2*, Eddison-Weskey, USA.

any moment. Moreover, even the knowledge itself about the rotation of the Earth around the Sun, and simultaneously about its rotation around its own axis, was not shared by the majority of people as recently as a few centuries ago, and this raised uncompromising discussions even among scientists.

Why does the famous illusion of living in the center of the Universe, at rest, raise all these furious debates in Galileo's time?

The answer is contained in the equivalence principle, which assumes that gravity and an acceleration of a frame of reference are indistinguishable *locally.*

In case of doubt, experiment has to decide on what is the truth. However, one has to perform a properly designed experiment and also one has to understand its results. Thus, let us mention that famous experiments, which seem to verify the rotation of the Earth, like those with the daily angular drift of Foucault's pendulums, confirm in some sense only the presence of an additional-to-gravity force, piercing our space and acting on all moving objects. We have already learnt that this is the Coriolis force.

Therefore, the unfortunate point of view of opponents of Galileo Galilei, who declined the reality of the Earth's rotation (both the Earth's rotation around its own axis and its rotation around the Sun), could be understood for their times, because they were motivated in part by the statement that the effects of rotation of the Earth are in fact equivalent to the effects of a special extra form of gravity which is always present in our world.

Nevertheless, as far as you put an appropriate question and address it in a properly designed experiment "It doesn't matter how beautiful your theory is, it doesn't matter how smart you are. If it doesn't agree with experiment, it's wrong." (Richard Feynman).

As long as gravity and an acceleration of a frame of reference are indistinguishable only locally, in order to analyze these illusions and to verify the truth, one has to propose some hypotheses on the form of the laws of gravity and to compare predictions following these laws with the results of *nonlocal* experimental observations like, for instance, those of laboratory measurements of gravitational gradients or astronomical observations of distant objects.

Thus, let us return to the idea of centrifuge in order to find out how our perception of flatness of space raises illusions, and how gravity enters into these illusions.

We will think of standing inside an *imaginary* tower of Pisa tower, which is a large hollow vertical cylinder with a radius R. Let it be nighttime, so we cannot see directly the shape of the walls in the darkness. Assume that we are standing very close to an internal wall of the tower. Thus we cannot immediately recognize that the wall is slightly curved; it looks to be flat to us (finally, it is a question of accuracy of your measurement/knowledge to define what is actually "flat" for you).

Imagine also that we are pushing a small ball (which we would consider to be shining and clearly visible in the darkness) with a velocity $\vec{u}$ in the direction along the floor and approximately parallel to the closest wall of the tower.

If the reader does not like a night adventure in an imaginary Pisa tower, they could also think of playing billiards in the darkness with a shining ball on a *round* table.

The shiny ball (in both these examples mentioned above) is initially found at a distance H measured from the closest wall of the tower. In order to describe the motion of the ball, we would like to introduce two "natural" variables: the distance l measured along the (slightly curved) wall and the distance r measured from the closest wall in the direction normal to it. At any moment of time the ball is characterized with a pair of coordinates $l(t), r(t)$.

Let us calculate how a trajectory of our ball would look like if expressed using these curved coordinates l, r.

Start from solving the Problem 1.25.

Problem 1.25. *Verify that a straight line, in particular a relatively short chord of a circle with a radius R, could be expressed in coordinates l, r as follows:*

$$r = H - \frac{l^2}{2R}. \tag{1.12}$$

Note: While deriving this equation, neglect second-order terms, which are proportional to H^2/R^2, and even smaller terms.

Yes, the ball trajectory, being expressed in coordinates r, l, is not a straight line. It is a *parabola*!

The reader knows that a parabola is a trajectory traced by an object, which falls freely at a constant acceleration in a spatially uniform gravitational field. Could we make this analogy between free motion described in curved coordinates and free fall in a gravitational field even closer? In order to analyze similarities and differences, let us estimate the value of an effective gravitational acceleration experienced by the ball in terms of parameters of the ball motion and the wall curvature.

Yes, it is easy to calculate the value of this acceleration as far as we take into account the approximate equality $l \approx ut$ (within the same level of accuracy):

$$r(t) = H - \frac{u^2}{R}\frac{t^2}{2}. \tag{1.13}$$

It is quite obvious from just looking at the general form of the above equation that the acceleration of the ball fall, measured in coordinates r, l, equals:

$$a = u^2/R. \tag{1.14}$$

One could easily recognize here a well-known expression for the centrifugal acceleration.

Solve the Exercise 1.14, which could also serve us as a third equivalent example to those with an imaginary Pisa tower and a round billiards table with a shiny ball.

Exercise 1.14. *Imagine that you are in a skiing resort, and decide to go skiing while the weather is still bad. The visibility is so poor that you see only the snow surface close to you and (nearly) nothing else. A cable car lifts you along a (nearly) straight line to the top of a mountain that (nearly) has a concave cylindrical shape (the cable-car line forms a chord with the mountain surface). While being in the cable car you feel that gravity is constant and the mountain surface is approaching towards you with acceleration. However, while later watching a film recorded from the car you get* *an impression that you are falling, together with the cable car, down to the mountain surface.*

Derive a relation between parameters of the problem (the cable-car's velocity, the curvature radius of the concave mountain surface, and the angular size of the chord formed by the cable-car line) which would correspond to the effective acceleration equal in the point of arrival to:

i) the standard gravitational acceleration,

ii) one tenth of this acceleration,

iii) one hundredth of this acceleration.

Could you point out some realistic sets of parameters of the problem, which meet conditions i), ii) and iii)?

Is there an evident indication in your film, which proves that it is not the funicular cabin that is falling down to the snow surface of the mountain but it is the snow surface of the mountain that is approaching with an acceleration?

Let us underline that the imaginary Pisa tower with an ice-skating rink is not rotating around its own axis, and the round billiards table in our example with the round edge and a shiny ball is not rotating around its symmetry axis, the mountain in the skiing resort not trembling or moving. Nevertheless we obviously observe the centrifugal acceleration in all these mentioned cases, even without rotating our frame of reference relative to an inertial one!

Why does, for instance, the ball in the imaginary Pisa tower, which is "in fact" moving freely along a straight line, seem to fall down to the tower wall with the effective acceleration u^2/R?

The answer to this question is evident: the illusion arises due to the special choice of coordinates, which are not straight but "curved". It is curious that we would not be able to realize that our coordinates are curved, if we were performing only truly local measurements. In contrast, if we were able to measure distances to other shining balls, for instance, then we would be able to judge whether the wall is curved, and thus to distinguish real and effective gravity forces.

And as soon as our objects and "frames of reference" are illuminated, thus we can perform further-than-local observations, we also can distinguish real and effective gravity forces.

The illusion of gravity, which arises due to the special choice of curved

Fig. 1.3   Fall up from centrifuge.

coordinates related to the railroad of Russian mountains in an amusement park in figure [1.3] is so strong that a bird, which has just escaped from a cage installed on a trolley, continues "falling up" for some time after the moment of escape. It feels the effect of artificial gravity stronger than the normal gravity. In fact, such artificial gravity is not an illusion as long as one cannot distinguish it from reality. An advanced reader could themselves formulate a problem related to the latter case.

Problem 1.26.   *Describe the motion of a cabin attached to a wheel in an attraction park. The wheel is rotating around its axis with the angular frequency w, the wheel radius equals R, and the arm of fixation of the cabin equals L.*

Note: Assume for simplicity that the friction and air resistance are negligibly small.

Note: Consider different regimes of motion, which correspond to different combinations of parameters of the problem.

Let us analyze more thoroughly how a curvature of our coordinates produces this effect of *artificial gravity* in terms of the mathematical formalism involved.

The distance between two close points in a plane is given in an expression, based on the Pythagorean theorem:

$$(ds)^2 = (dx)^2 + (dy)^2. \tag{1.15}$$

One could ask a question based on this definition: What is the shape of a curve, that provides the shortest path between two points with coordinates (x_a, y_a) and (x_b, y_b) in the plane? It is well known that this "curve" is in fact simply a straight line. In order to prove this statement rigorously, one should first write down a general expression for the path length measured along an arbitrary curve $y = y(x)$ drown between these two points in the plane:

$$s = \int_{x_a,y_a}^{x_b,y_b} ds = \int_{x_a,y_a}^{x_b,y_b} \sqrt{(dy/dx)^2 + 1}\,dx. \tag{1.16}$$

For simplicity this expression is written for only two spatial dimensions x, y; however, this approach could be easily generalized to a larger number of spatial dimensions.

One should compare all possible (so-called *virtual*) curves $y = y(x)$ passing through the given points (x_a, y_a) and (x_b, y_b), and calculate for what such a curve the integral (1.16) with the meaning of the path length gains its minimum value. An efficient way of solving such problems consists of using the method of *variational calculus*.

The general procedure of this mathematical theory is based on the fact that the curve $y(x)$, which minimizes (or maximizes) the integral $\int_{x_a,y_a}^{x_b,y_b} L(y(x), y'(x))dx$, could be calculated by means of solving the following differential equation:

$$\frac{dL}{dy} - \frac{d}{dx}\frac{dL}{dy'} = 0. \tag{1.17}$$

The *L*-function in the above expression is called the *Lagrange function*[14]. In our case of a curve with the minimum length between two points in a plane, the Lagrange function is equal $L(y', y) = \sqrt{(dy/dx)^2 + 1}$, and the equation (1.17), resolved relative to the function $y(x)$, has the following simple form:

$$\frac{d^2y}{dx^2} = 0. \tag{1.18}$$

This condition of zero curvature corresponds simply to an equation, which describes an arbitrary straight line $y(x) = Cx + D$ defined in coordinates x, y.

If the line passes through two points, (x_a, y_a) and (x_b, y_b), then the constants C and D would be uniquely defined, so that $y = y_a + \frac{(y_b - y_a)}{x_b - x_a}(x - x_a)$.

We have performed this exercise in order to point out that after departing from the interval between two close points in a plane (which is defined as follows):

$$ds^2 = dx^2 + dy^2, \tag{1.19}$$

we conclude that the shortest path between two arbitrary points in the space with the above-given form of interval is along a curve $y = Cx + D$, i.e. along a straight line.

This result concerning a straight line as the shortest path between two points in a plane is an expected one. Would the result be more complex if we decide to describe the world not in flat coordinates but in slightly curved coordinates l, r, which we used previously for describing the ball motion near a curved wall?

(*Note*: This approach is not as exotic and unnatural as it seems to be. Imagine that you have not yet learnt that the Earth is not flat but round. Then you naturally tend to trace a coordinate grid on the round surface of the Earth while supposing that it is actually flat. And thus you are going to "naturally" get all kinds of artificial effective forces.)

---

[14]Lagrange, J.-L. (1811). *Mecanique Analytique, Courcier*, Cambridge University Press, 2009.

One could establish that an interval between two close points expressed in coordinates l, r is:

$$(ds)^2 = (dr)^2 + (1 - r/R)^2 (dl)^2 \approx (dr)^2 + (1 - 2r/R)(dl)^2. \qquad (1.20)$$

A curvature of space, being defined in such a way, manifests itself in a small correction to the expression (1.19) for the distance between two arbitrary points; this correction is contained in the extra term $-2r/R$. (This result is not specific for the two-dimensional representation used above; a similar result is also valid for other numbers of spatial dimensions.)

Although this correction $(-2r/R)$ is small, nevertheless it is large enough to "turn a straight line into a parabola", and thus to produce the full illusion of the effect of gravity!

Problem 1.27. *Follow the rules of variational calculus described above and verify that the expression (1.20) for the shortest curve expressed in our (l, r) coordinates would look like equation (1.12), which describes a chord of a circle.*

Thus we conclude that a free object moves in both cases – in spaces with flat and curved coordinates – along a curve of the minimum length. However, in the case of Euclidian (plane) space such a curve is a straight line, while in a curved space, which "produces" artificially the effect of gravity, such a curve is no longer a straight line – it might become a parabola, but it could also get a much more complex shape if only the curvature of our coordinates were more complicated.

Finally, what is curved in reality – coordinates or space? The answer might be also quite relative... Saying that, we could repeat after Albert Einstein that *gravity is a curvature of our space*; and also it is a *curvature of our time* as long as we take into account the relativistic effects. In such a world, objects do not fall down in a gravitational field, instead space itself is curved in such a way that we might think that relative distances change nonuniformly.

So gravity is an illusion of curved space!

In order to enjoy this close relation between gravity and a curvature of space, let us return for a moment to the closely related problem with a shiny ball, which slides with no friction in the darkness on the ice-skating rink in the vicinity of a wall of the imaginary Pisa tower. We assume that the reader has already realized that the ball in our thinking experiment would bounce back and forth, thus moving along the surface of the cylindrical wall.

The general picture of the ball motion, which is represented in our curved coordinates (l, r), would be equivalent to the picture of the motion of a ball, which is thrown parallel to the surface of the Earth (with the gravity intensity on the surface of the Earth being equal to $a = u^2/R$), and which then follows a parabolic trajectory until it hits the surface and gets reflected. With no energy dissipation, such a periodic motion would continue infinitely.

It is almost the same bouncy ball as it was in our very first examples illustrated, for instance, in figure [1.1]. However, now the ball exhibits an extra motion along the surface, i.e. it has an additional component of velocity, which is parallel to the surface. Besides, there is also the following difference between the two kinds of bouncy balls, which complements their certain similarity.

While the gravitational force in the first case is the same for all balls with any value of their horizontal velocity component (this observation would be valid as long as the ball velocity were much smaller than the velocity of light), the strength of artificial gravity in the second case changes as a function of the value of *longitudinal velocity component*. This effect is stronger for larger longitudinal velocities, and it is canceled out to produce a kind of "levitation" for a zero-velocity component parallel to the surface.

Problem 1.28. *Calculate the distance H measured from the wall of a cylinder with a radius R in our above-described example of a shiny ball in an imaginary Pisa tower, such that the ball thrown parallel to the wall would return exactly to the same point after making a full circuit?*

Problem 1.29. *You are enjoying a rotating carousel with a radius R and a linear velocity u at your point. You look towards the center of the carousel and put your arm forward in the same direction. Then you release a small ball from your hand.*

Calculate the moment of time when the ball would reach your chest, assuming the length of your arm is H.

The example, which we have discussed, underlines the *duality* between two following problems: the problem of the falling of an object to a plane surface with a given acceleration of the fall (the acceleration is not necessarily a constant) and the purely geometrical problem of crossing a straight line with a curved surface (the curved surface is not necessarily a sphere or a cylinder).

In particular, we see the relation between the problem of a ball bouncing in the gravitational field of the Earth and the problem of crossing a circle with a straight line.

The deep sense behind this analogy is contained in the equivalence principle, which represents a gravitational interaction in terms of spatial geometry. This geometrical representation of a gravitational interaction is possible only due to its *universal character* (the universal character understood in the sense that gravity acts on all objects in a similar way), as well as due to the fact that curvature of space also provides equal effect on all objects independent of the value of their mass.

Exercise 1.15. *A ball is thrown parallel to the surface of the Earth. It is found initially at a height H above the surface, and is going to hit the surface at a distance L from the initial position.*

Calculate the initial velocity of the ball.

Solve this problem using its duality with the problem of a circle crossing a chord.

Exercise 1.16. *Calculate the acceleration of a ball moving along a chord of a cylinder at a certain velocity, using the law of energy conservation, written in cylindrical coordinates.*

Exercise 1.17. *A relative distance r from a circle to a straight line, which*

we have just introduced in our example, changes as the square of the distance l, $r = H - l^2/(2R)$ measured along the circle, provided we neglect
terms like H^2/R^2 and even smaller terms.

Find the shape of a curve, for which this law would be satisfied with
higher accuracy.

Exercise 1.18. *Imagine that you are walking through a dark straight se-
cret underground tunnel, which connects two points on the surface of the
Earth. The angle between the gravity force vector and the direction of your
walk is different in different points of the tunnel, so the straight tunnel
would seem curved to you.*

*Find the illusive form of the tunnel, which your reception would dictate
to you.*

Problem 1.30. *Imagine that you make experiments with a cone funnel,
which is aligned vertically and its wider part is directed upright. The funnel
is characterized by an angle α and a height H. You throw a small ball along
the inner surface of the cone with an initial velocity vector set in the plane
orthogonal to the cone axis. The ball could slide with no friction along the
cone surface.*

*Find the range of absolute velocity values, for which the ball would stay
trapped inside the funnel.*

To summarize our findings, we have understood that free objects move
along the shortest straight lines in Euclidian space, which is free of gravitational potentials. But in the presence of a gravitational field the objects
would move along the shortest line in a curved space (such a curved line
would be a parabola in our example with a constant gravitational potential,
or it would be a curve with a more complex shape in the case of a more
complex form of gravitational potential).

One can ask a question: Could we somehow extend this statement to
all kinds of interaction?

Another question: Is it so that all objects always choose a "shortest
way" in some sense?

A long time ago Pierre de Fermat formulated an analogous principle[15] for the propagation of light. He found that light moves always between any two points along such a line that the time of passage τ between these two points is minimal:

$$\tau = \int_a^b ds/c. \tag{1.21}$$

The speed of light c is constant in empty space, so the *Fermat principle* is equivalent to the statement that light moves along a straight line in empty space.

Activities aiming at the generalization of the "shortest path" approach, i.e. at the identification of the most universal analogues statement, which would be also valid for all objects participating in the interactions of any kind, finally motivated the formulation of the so-called principle of *least action*[16]. This principle states that you could always define such an integral that it would play the role of the "shortest path" for all systems and interactions.

The principle of least action, in its most universal form, states that all objects move from a given moment of time t_1 to another moment of time t_2 along such a trajectory $\vec{q}(t)$ that the effective shortest path, the *action* S gets its minimum value:

$$S = \int_a^b L(\vec{q}(t), d\vec{q}(t)/dt))dt. \tag{1.22}$$

It is amazing that this short formulation includes the whole of classical physics!

A novel issue in the latter consideration is that the least-action approach analyzes not only a single trajectory but also other different virtual trajectories (in a general case, an infinite number of such virtual trajectories) between given points along which the object could evolve. It states

---

[15]This principle was stated by Fermat in a letter dated January 1, 1662, to Cureau de la Chambre. For the history of this question see Mahoney, M.S., *The Mathematical Career of Pierre de Fermat, 1601-1665* 2nd edition, Princeton University Press, 1994, p. 401.
[16]Landau, L.D., and Lifshitz, E.M. (1969). *Mechanics, A Course of Theoretical Physics, Vol. 1*, Pergamon Press.

that only those trajectories among all possible virtual ones become *physical*, which minimize the action. This approach has a deep relation to quantum mechanics, as we are going to see below.

However, the principle of least action has a certain meaning and thus it allows us to perform any concrete calculations for a given system participating in an interaction of a given type, only in the case the Lagrange function L is established for this system. We have already mentioned that the Lagrange function determines the differential equation, which allows us to calculate the physical trajectory (1.17). Let us look more thoroughly at the form of the Lagrange function.

One could notice, after having analyzed the definition of the Lagrange function given in equation (1.17), that this equation includes only derivatives of the Lagrange function, and thus this very important function is not uniquely defined. In particular, if one adds to a solution of equation (1.17) another arbitrary function, which is a full time derivative of another function of coordinates and time, the action S would change only by a constant value.

Problem 1.31. *Check for yourself the latter statement (that the Lagrange function is not uniquely defined and that the action S would change only by a constant value if one adds an arbitrary function, which is a full time derivative of another function of coordinates and time, to a solution of the equation (1.17), which defines the physical trajectory).*

As a derivative of a constant value is simply equal to zero, then this change of the action by a constant value would not affect the choice of the physical trajectory, which is calculated in such a way that it minimizes the action. Therefore, this significant freedom of defining ambiguously the Lagrange function, without affecting the physical trajectory, results in important consequences within the quantum description of motion, which we are going to appreciate in the next chapter.

However, instead of passing immediately to the next (*quantum*) level of description of motion, we would like to analyze here in more detail the classical Lagrange function. In particular, how could the classical Lagrange function be established for various systems? It would be very instructive to investigate this procedure, departing from first principles, for instance in a simple case of free motion described in an inertial frame of reference.

First, we remind the reader that empty space in an inertial frame of reference is *homogeneous*, and no time instant is unique (i.e. time is *homogeneous* as well). Departing from these rather general properties of space and time, we could make quite concrete conclusions on the allowed or not-allowed form of the Lagrange function for any particular case. Namely, this statement means that the Lagrange function L cannot depend explicitly on coordinates or time.

Second, as far as free space in inertial frames of reference is also *isotropic*, then all directions in space are equivalent. Thus the Lagrange function could be "constructed" only using such physical values, which are not modified while the frame of reference is rotated by an arbitrary angle. Therefore, we conclude that the Lagrange function L is a function of v^2 only, where v is the velocity of a freely moving object in the given inertial frame of reference.

Third, the Lagrange function should obey the principle of relativity. This statement imposes that the Lagrange function L in another inertial frame of reference, which is moving at a velocity $\vec{v}_0$ relative to our frame of reference, should change only by a full time derivative of an arbitrary function of coordinates and time. For small values of velocity v_0 (neglecting here the relativistic effects) we could thus write the following equation for the transformation of coordinates:

$$L(v'^2) = L((\vec{v} + \vec{v}_0)^2) = L(v^2) + dL/d(v^2)(2\vec{v}\vec{v}_0). \qquad (1.23)$$

The principle of relativity imposes that a change of the Lagrange function L is a full time derivative of an arbitrary function of coordinates and time. Thus:

$$\frac{dL}{d(v^2)}(2\vec{v}\vec{v}_0) = \frac{df(v, \vec{r}, t)}{dt}. \qquad (1.24)$$

This is possible only if:

$$L(v^2) = const \times v^2. \qquad (1.25)$$

The above equation (1.25) is nearly the final form of solution for the Lagrange function, which is usually written for free motion of an object in an inertial frame of reference, as follows:

$$L = \frac{m}{2} v^2. \qquad (1.26)$$

Here m is the *inertial* mass of the particle.

Problem 1.32. *Using the principle of least action, verify that the Lagrange function defined in equation (1.26) describes the motion of an object along a straight line.*

Compare the action for a free object to the length of a curve between two points (1.16).

Problem 1.33. *Verify that the action for a freely falling particle with the initial velocity v_0, aligned along the vertical direction, equals:*

$$S = mt(v_0^2/2 - v_0 g t + g^2 t^3/3). \qquad (1.27)$$

In a special case of a system, which is isolated from the external environment, with the total energy E of the system being conserved, the least action principle could be formulated in a form, which is exactly equivalent to our approach of the "shortest path". Namely, this principle establishes that a physical object should move along such a trajectory between two given points *in the space* with coordinates q_1 and q_2, that the following action,

$$S = \int_{q_1}^{q_2} p(q)dq - E(t_2 - t_1), \qquad (1.28)$$

gets its minimum value. Here $p(q)$ is the momentum of the object. It could be expressed via the total energy E and the potential energy $U(q)$ in a given point q as follows: $p(q) = \sqrt{2m(E - U(q))}$.

In the case of free motion the momentum is constant $p(q) = \sqrt{2mE} = const$ and the action S approaches its minimum value, while the integration is performed along a straight line, which connects the initial point and the final point. An interaction would transform the shortest path from a straight line into a certain curve, and the momentum would no longer be constant along this curve.

We see that the special role of inertial frames of reference consists of distinguishing the uniform motion of a freely moving object along a straight line and an accelerated motion of the object, which could be conditioned only by the gravitational interaction with other objects. This role relies on the concept of empty, plane, homogeneous and isotropic space. Geometric properties of our space are essential here, namely the definition of a straight line in our space.

The equivalence principle prompts us to "explain" the universal character of gravity in terms of a curvature of space-time. It states that an accelerated motion of an object due to gravity is a sort of illusion in our curved space-time. It also states that all objects in a gravitational field move along "shortest paths" like in the case of free motion. However, these shortest paths in the presence of a gravitational field are determined by the geometry (*metric*) of curved space.

Thus, accelerated frames of reference have "equal rights" with inertial frames of reference, while the motion in a gravitational field has "equal rights" with free motion.

1.4 Συνοψις

We started the book with a reflection on such a simple phenomenon as the *free fall* of an object in the gravitational field of the Earth.

It allowed us to introduce the concept of *physical experiment*, which is characterized in particular by a *principal phenomenon* to be studied and *false effects* to be eliminated, suppressed or accounted for.

Distinction between a principle phenomenon and false effects resulted in a first guess on the concept of *empty space* in classical mechanics.

A key feature of a properly designed physical experiment is its *precision*, and thus *reliability* of results and conclusions.

We continued with a treatise on an elastic *bouncy ball*; we optimized various experimental conditions, suppressed false effects and underlined such a convenient feature of a bouncy ball experiment as (approximate) *periodicity* of the ball motion.

The periodicity and *long time of observation* make experiments with such bouncing objects a useful *instrument* for various *applications*.

Inspired by Galileo Galilei, we found that the bouncing period does not change as a function of the ball mass value:

$$T = 2\sqrt{\frac{2H}{g}}. \tag{1.29}$$

The period T is a function of only the initial dropping height H and the gravity strength g for *any object*.

This finding is amazing, and thus we decided to explore to what degree it is *universal*.

In particular we tried to extend it to a more general case of objects moving arbitrarily in a gravitational field and to a case of objects interacting via *other-than-gravity forces*.

However we promptly understood that as soon as multiple degrees of freedom of a complex system are involved, motion starts to be complicated, even in the case of only gravitational forces. Thus, while the conclusion on mass-independence of motion is valid for *accelerations* of point-like bodies, it is not automatically valid for integral characteristics of motion of a complex system such as a motion *period*. Other-than-gravity interactions are an essential part of any complex system. Generally speaking, their account violates the statement of mass independence of motion in a gravitational field of any realistic macroscopic object.

All these useful observations are encoded in *Newton's equation of motion*:

$$m_i = M_i \sum_{i \neq j} M_j \frac{(\vec{r}_j - \vec{r}_i)}{|(\vec{r}_j - \vec{r}_i)|^3}. \tag{1.30}$$

To what degree is the statement on mass-independence of accelerations accurate and how can it be verified experimentally?

When considering the *accuracy* on an eventual experimental verification of the mass-independence of accelerations, we promptly understood that Newton's equation (1.30) is formulated for *material points* with no size, while real objects have finite spatial extensions. Nevertheless, as far as gravity is approximately *homogeneous* within the size of the object, the corresponding equations are approximately correct for the motion of its *center of mass*.

Thus we concluded that the universal statement on the mass-independence of accelerations is valid with a good accuracy only for objects with spatial extensions negligibly small compared to the scale of inhomogeneity of a gravitational field.

Another practical question, which we have to address while verifying our finding experimentally, is: Relative to what *frame of reference* should we measure accelerations? Would a result depend on a choice of a reference frame? Isn't it so that we would observe different accelerations as a function of choice of a frame of reference?

At this point we returned to Galileo Galilei, who formulated the *principle of relativity*, which states that there are special, inertial frames of reference among numerous other frames. While we study properties of empty space in inertial frames, we will find that *space is plane, homogeneous and isotropic*.

We can define *universal coordinates* in such frames of reference; they could be physically realized, for instance with rigid rods. We can define a *straight line* in a universal manner as a line of propagation of light. We can define universal time measured using a *clock* made of two parallel mirrors with a light pulse bouncing between them.

As soon as we had done so, we returned to formulating physical laws in such inertial frames of reference. The principle of relativity states that all physical laws should *look the same* in inertial reference frames moving relative to each other at a constant velocity. The acceleration of any object relative to these frames of reference is the same just as we want in order to provide the universal formulation of the independence of acceleration as a function of the mass in a gravitational field.

Thus before performing a stringent experimental test we have to verify that our frame of reference is an inertial one. How could we do that practically?

We studied several examples and found out that extra *Coriolis-like effects* appear in rotating frames of reference, which are thus not inertial ones; and that as long as we are enclosed, for instance in a limited space of an elevator, we can hardly distinguish gravity and a translational acceleration of the elevator (the acceleration of our frame of reference). "Can hardly" would be replaced by "cannot" in the previous statement if we were able to perform a truly *local* experiment measuring gravity and acceleration on a single point.

In fact, as far as we are able to measure a gradient of the force in order to verify the gravity effect, we would always distinguish gravity and acceleration. The main reason, which explains why a measurement of *gradient force* is difficult in usual laboratory conditions, consists of extreme weakness of a gravitational interaction and, as a result, of huge characteristic spatial scales of changing the gravitational field compared to sizes of an elevator, of a ship, of a supersonic jet or a bouncy ball.

However, the *hypothesis of nondistinguishability* of gravity and acceleration was an important starting guess.

With the reservations mentioned above, especially those about the distinction between *local* and *nonlocal*, or *global*, observations, we could say that in fact we cannot distinguish gravity and a *curvature of our space-time*. Indeed, we studied an example in which slightly curved coordinates were used, and the effect of gravity *emerged* with a good precision there only due to this curvature of coordinates.

We concluded that the most universal statement, which we could make from our analysis and which would contain all the facts mentioned above, is the statement on *equality* of the inertial mass m and the gravitational mass M of an infinitely *small* particle:

$$m = M. \tag{1.31}$$

This is the content of the *principle of equivalence*, the cornerstone of general theory of relativity, one of the most beautiful and intriguing physical theories, which understands gravity in terms of curvature of space-time caused by the presence of matter in our Universe.

But does this statement mean that we could not distinguish between an inertial frame of reference and an accelerated one? If so, inertial frames of reference would be useless!

An answer to this question involves *global* observations. Global observations could be of a varied nature – they could involve, for instance, a measurement of *inhomogeneity* of a gravitational field (such as its gradient and higher order derivatives) performed in a *very limited volume* of space, or they could also consist of measurements, which involve observations of distant objects.

We could test geometrical properties of *empty* space (i.e. of a region of space that is located far away from any massive object). We remember that a straight line is defined using a light ray. If we measure a sum of the internal angles in a triangle defined by means of light rays and compare it to π, we could conclude then on whether our space is flat or curved. In an inertial frame of reference we should get π; this result would indicate that physical properties of our empty space correspond to the Euclidian geometry.

The idea that motion occurs along a curve with the *shortest length* (such a curve is a straight line in the Euclidian space, and it is a parabola in the above-mentioned example of a cylindrical wall in the darkness) results in introducing the *principle of least action*. This principle states that among all possible *virtual* trajectories only those trajectories that minimize the action become real. It is curious that this principle includes all of *classical physics*!

However, we are still not completely satisfied at this point. We would like to know whether the equivalence principle, the least action principle and the principle of relativity would be valid not only in classical mechanics but also in the world of tiny particles, such as atoms, electrons, neutrons, etc or more precisely in the world of *quantum mechanics*, which includes macro-scopic quantum-mechanical phenomena (not, however, studied in depth in this book).

Our major concern here is about the *concept of coordinates* as an idealization of a rigid body and about the concept of a "trajectory", which necessitate the concept of coordinates. If microscopic particles are waves, as stated in quantum mechanics, if there is no more place for a trajectory of a tiny particle, how could we then talk about straight lines (or curves) in the world, deprived of rigid coordinates?

How could we formulate the above-mentioned fundamental principles, based on those concepts, in the micro-world?

Is it possible to *combine* quantum mechanics, the principle of relativity and the equivalence principle?

These are the intriguing questions we are going to discuss next.

Chapter 2

When a Particle Becomes a Wave

2.1 Quantum Fall

"If, in some cataclysm, all of scientific knowledge were to be destroyed, and only one sentence passed on to the next generation of creatures, what statement would contain the most information in the fewest words? I believe it is the atomic hypothesis that all things are made of atoms – little particles that move around in perpetual motion..."

Richard Feynman estimated the value of atomistic theory as this important. The concept displaced the illusion of matter being a continuous medium. This illusion had been strongly supported by everyday experience, which prompts that an under-pressured domain is rapidly filled in from the environment, whether it is gas, liquid or something else. The great power and great temptation of atomistic theory consists of its promise to explain all the manifold of phenomena in our world in terms of simple motions of little particles, which obey the laws of *Newtonian mechanics* that are well established for macroscopic objects.

It is surprising these days that this concept was not widely accepted even by scientists as recently as the early twentieth century. It is not widely known that Albert Einstein, whose name is rather associated with special and general theories of relativity, contributed a lot to proving this theory by means of his treatise on the Brownian motion. One problem in accepting atomistic theory was simple: nobody had ever observed atoms.

However, there is a deeper reason for concern, namely the uncertainty on whether one could reasonably assume that such small particles, which we can not even see, obey the same laws of motion as macroscopic objects do. A first natural response to this question would consist of our intuition that the world is expected to be constructed in the same manner on all spatial scales, large and small. But this intuition also appeared to be a great illusion!

An attempt to simultaneously use Newtonian mechanics and electro-dynamics (the two most elaborated theories in physics at that time), for explaining the behavior of atoms, failed. Such a failure appeared to be even more striking for electrons, which are the lightest constituents of atoms. Soon it was realized that a new, quantum, mechanics based on new and counterintuitive principles, is required to displace our illusions concerning the world of tiny particles.

There are two basic statements of quantum mechanics that make it a theory which "nobody understands" according to the famous observation of Richard Feynman.

The first statement of quantum theory is that *all particles are in fact waves*. It means, in practice, that particles, at least those with small enough mass, at certain conditions manifest properties rather typical for waves.[1] A characteristic feature of such wave-properties consists of the phenomenon of *interference*, or the ability of such waves to enhance each other and (what is even more surprising!) to suppress each other in certain places.

This wave property is characteristic not only for material particles, it has been well known in light optics. If light passes through a narrow slit, one could observe a sequence of bright and dark parallel stripes on a screen behind the slit, provided that the size of the slit is comparable with the wavelength of light. This interference picture of the light-intensity stripes could be explained and precisely described analytically, as soon as the wave nature of light is accepted as a starting point.

While an experiment of this type was performed also with electrons (and later analogous measurements were done with most other elementary particles available, as well as with atoms and molecules) then undoubtable interference phenomena were observed as well. This is the main content of the statement that particles are waves. No one knows any other self-consistent explanation of these experiments, alternative to the idea of the wave nature of particles.

The second statement of quantum theory, which makes the whole picture of the motion of tiny particles even more contradictory to common sense and everyday experience, is that such particles, being waves, still *keep their corpuscular nature simultaneously*. In particular, once they are observed, they always appear in the detector entirely with a given charge, mass, and

---

[1]de Broglie, L. (1924). Recherches sur la théorie dés quanta, *Thesis, Paris*; (1925) *Annal. Phys.* **3**, p. 22.

so on. This property means that a microscopic particle could not be thought of as a cloud, which fills space and transmits some kind of wave.

The crucial expression, which relates a momentum of a *particle* p and a corresponding spatial wavelength λ of a corresponding *wave* at a given instant of time is:

$$p\lambda = 2\pi\hbar. \tag{2.1}$$

The relation between the energy of a *particle* and the temporal period T of a corresponding *wave* in a given spatial point is:

$$ET = 2\pi\hbar. \tag{2.2}$$

A fundamental physical constant $\hbar$, which enters in these two expressions, is the same. It is called the *Planck constant* and equals about $6.6 \cdot 10^{-16} eV \times s$. It is a fundamental value, which measures quantum features of the *wave-particle*. It also defines the fundamental scale, at which the wave properties of particles start to be essential or dominant. In the following we are going to return to these fundamental expressions. As you could see, they establish a new type of *relations between space and time*.

How could these particle and wave properties coexist simultaneously? Is it possible to propose a self-consistent formal description which combines these two contradicting descriptions? Yes, it is possible and the wave-particle *dualism* addresses this challenge. It assumes that one associates with a particle a *function*, which is defined *in any point of space at any moment of time*, the so-called *wave-function* $\Psi(\vec{r}, t)$, with properties described and commented on below.

Spatial and temporal properties of this wave-function are directly related to the momentum and energy of the particle.

Each wave-function describes a certain physical *state* of a particle; that is why the space of such wave-functions (also known as the *Hilbert space*) turns out to be a very important concept in this context. This is the wave-side of the story.

The particle side of the story is given in a simple rule: The modulus square of the wave-function $|\Psi(\vec{r}, t)|^2$ calculates the *probability density* of a particle observation in a given place at a given time.

It is not easy to consolidate these counterintuitive statements with our everyday experience and common sense. If the reader could do that, they are probably the only ones who understand quantum mechanics. Numerous attempts to discover a more "classical" interpretation of quantum phenomena, in particular those based on so-called "hidden parameters" of motion, failed. This is a signature of a really deep truth: it contradicts intuition and is hardly believable.

(*Note*: Such theories are called theories with "hidden parameters" of motion. They were proposed in order to explain quantum phenomena using a concept of "*insufficient knowledge*", in analogy, for instance, with the stochastic behavior of a dice, which could be, in principle, predicted precisely provided we have at our disposal an absolutely precise knowledge of all its initial conditions. Albert Einstein considered this hypothesis; his skepticism about quantum mechanics is concentrated in his well-known phrase: "God does not play dice". Perhaps a reason for Einstein's nonacceptance of quantum mechanical concepts was his provision about the difficulty combining it with the general theory of relativity.)

"If you thought that science was certain – well, that is just an error on your part" (Richard Feynman). In order to check your quantum intuition try to solve the following problem.

Problem 2.1. *Using only the expressions $p\lambda = 2\pi\hbar$ and $ET = 2\pi\hbar$, guess what the wave-function of a freely moving particle with a momentum $\vec{p}$ would look like.*

To continue, we have to immediately write down a solution for Problem 2.1:

$$\Psi_p(\vec{r}, t) = \exp(-iE_p t/\hbar + i\vec{p}\vec{r}/\hbar). \tag{2.3}$$

As long as the wave function of *free* motion $\Psi_p(\vec{r}, t)$ is defined, the momentum of a particle can be calculated using the following straightforward procedure:

$$-i\hbar\vec{\nabla}\Psi_p(\vec{r},t) = \vec{p}\Psi_p(\vec{r},t). \tag{2.4}$$

This procedure is important as it relates the momentum of a *particle* and the spatial structure of a *wave*. We are going to use this momentum–space relation further on.

The practical advice of quantum mechanics for any concrete problem is to solve a *wave-equation* in order to find a wave-function, then to predict the *probability* of measuring any physical value of the particle using this wave-function.

The famous *Schrödinger equation*[2], the quantum equation of motion, allows us to calculate a wave-function in any point of space at any moment of time, and thus it provides us with the full description of a quantum system.

In order to get a feeling how such an equation could be discovered in a more general case, try to derive it yourself by means of solving Problem 2.2.

Problem 2.2. *Guess what the form of equation is for the wave-function of a freely moving particle.*

Note: Take into account that the solution of this equation is given in the expression $\Psi_p(\vec{r},t) = \exp(-iE_pt/\hbar + i\vec{p}\vec{r}/\hbar)$ *(2.3).*

Note: Take also into account that the equation, which has to be found, should not change while being written in a rotated frame of reference as it should follow from the invariance of physical laws under the operation of rotation.

Again, we write down the solution for this problem immediately as we need it to have known explicitly in order to continue:

$$i\hbar\frac{\partial\Psi(z,t)}{\partial t} = -\hbar^2\frac{1}{2m}\frac{d^2}{dz^2}\Psi(z,t). \tag{2.5}$$

---

[2]Schrödinger, E. (1926). Quantisierung als Eigenwertproblem (Zweite Mitteilung), *Annal. Phys.* **384**, p. 489.

Using this very brief outline of principal statements of quantum mechanics, we underline that the wave character of motion of a quantum particle imposes that the particle has to be associated with a wave-function, which (generally speaking) should be defined in every point of space at any moment of time. Such a wave-function provides a complete self-consistent description of quantum motion.

In order to understand what is in fact motion in quantum case, as well as to analyze what physical characteristics of quantum motion could be measured, let us think of another "imaginary Pisa tower" experiment, but now with a falling object in the form of a quantum particle. At the moment we do not care what kind of physical particle it is, although below we are going to discuss its eventual nature in detail, with references to real realizations and experiments.

Now it is time to write down the Schrödinger equation for a *falling quantum object* and to analyze its wave-function. In order to get such a wave-function explicitly, let us solve this equation for a particle, which falls down in the gravitational field of the Earth only in the vertical direction z. This equation is different from the Schrödinger equation (2.5) for a freely moving particle by an additional potential term Mgz introduced in its right side; it looks like:

$$i\hbar \frac{\partial \Psi(z,t)}{\partial t} = \left(-\hbar^2 \frac{1}{2m} \frac{d^2}{dz^2} + Mgz \right) \Psi(z,t). \qquad (2.6)$$

In the above equation we distinguish the inertial mass m and the gravitational mass M in order to be able to treat the *equivalence principle in quantum mechanics*.

At this point we would like to address to an attentive reader who has probably noticed that these two masses enter asymmetrically in the Schrödinger equation (2.6).

Exercise 2.1. *Verify whether the Schrödinger equation* $i\hbar\frac{\partial \Psi(z,t)}{\partial t} = \left(-\hbar^2 \frac{1}{2m} \frac{d^2}{dz^2} + Mgz \right) \Psi(z,t)$ *(2.6) for a freely falling particle could be written in such a form that the gravitational mass M and/or the inertial mass m are reduced.*

As you could conclude after thinking about Exercise 2.1, the masses cannot be reduced, and thus the wave-equation and its solutions are functions of both values m and M.

There are paradoxes hidden in equation (2.6).

First, the statement that gravitational and inertial masses could not be reduced in the Schrödinger equation for the quantum fall, sounds very suspicious: it looks like the equivalence principle were violated in quantum mechanics. At least it sounds violated in the sense that the wave-function, or quantum motion, of a particle in a gravitational field does change as a function of its mass in a general case. At least it is indeed violated if understood in such a naive classical sense!

If we would like to insist on keeping the equivalence principle in the quantum world, should we try to redefine it? Is this at all possible?

Before discussing these questions, let us derive an explicit solution for the Schrödinger equation (2.6). In order to solve uniquely this equation, we have to define the initial conditions and the boundary conditions.

First, we are going to study a special solution corresponding to the classical idea of a point-like particle (with no size). Namely, we assume that the height of a particle $z = z'$ is *precisely known* at a time instant $t = 0$.

This assumption corresponds to the wave-function, which is equal zero everywhere, except for the given height $z = z'$ (in the following we are going to consider "one-dimensional" motion along the vertical coordinate z in a gravitational field, and we are going to momentarily ignore all motions in other directions, as in the classical case).

Such a wave-function is given in the distribution function, known as the "Dirac delta":

$$\Psi(z, t = 0) = \delta(z - z'). \qquad (2.7)$$

It is a mathematical expression of the idea of a point-like source.
(*Note*: The main property of the delta function is given in the following

expression:

$$\int_{-\infty}^{+\infty} \delta(z - z')f(z)dz = f(z') \tag{2.8}$$

with an arbitrary function $f(z)$.)

   The solution we are looking for now is of special importance because it is going to serve us for calculating other solutions of more complex problems. Therefore, we denote this function with a separate capital letter $G(z, z', t)$ (it is called the Green function[3]). A modulus square of this function $|G(z, z', t)|^2$ equals to the probability density to observe a particle in the vicinity of a height z at a time t, provided that at the initial moment $t = 0$ the particle was localized at the height $z = z'$.

   As soon as this function $G(z, z', t)$ is known, one can use it to "construct" a solution corresponding to any other initial distribution $\Psi_0(x)$, as follows:

$$\Psi(z, t) = \int_{-\infty}^{+\infty} G(z, z', t)\Psi_0(z', t = 0)dz'. \tag{2.9}$$

Now we can write down the function $G(z, z', t)$ explicitly:

$$G(z, z', t) = \sqrt{\frac{m}{2\pi i \hbar t}} \exp\left(iS(z, z', t)/\hbar\right), \tag{2.10}$$

$$S(z, z', t) = \frac{m}{2t}(z - z' - \frac{Mgt^2}{2m})^2 + Mgz't - \frac{M^2g^2t^3}{6m}. \tag{2.11}$$

Let us first use this solution for calculating what would happen to the particle an instant after the moment of its release. According to quantum mechanical rules, we have to estimate the probability density to observe a particle (which was found initially at a height z') in a small vicinity of a new height z at a time t:

$$dP(z, z') = |G(z, z', t)|^2 dz = \frac{m}{2\pi \hbar t}dz. \tag{2.12}$$

---

[3] For a discussion of Green functions, as well as other issues in quantum mechanics, see Baz', A.I., Zel'dovich, Ya.B., and Perelomov, A.M. (1969). *Scattering, Reactions and Decay in Nonrelativistic Quantum Mechanics*, Jerusalem, Israel Program for Scientific Translations.

This is a remarkable and counterintuitive result! It shows that the probability density to observe an initially *precisely localized* particle at any following instant of time is *the same at any height*. In other words, the particle could be observed everywhere after an infinitely short period of time, provided it has been initially localized inside a tiny volume and then released. In particular, it could be observed higher than its initial height. Would a quantum particle fall up?

Another direct consequence of the above statement is that the velocity of quantum motion of a particle (which is understood as the velocity of spatial spreading of its wave-function) could become arbitrarily large. In particular, it could be even larger than the velocity of light. Moreover, it could be directed both upwards or downwards. Thus, for instance, the quantum particle could fall upwards with the velocity larger than the velocity of light... This is something very suspicious.

Exercise 2.2. *Verify explicitly whether the equation $dP(z, z') = |G(z, z', t)|^2 dz = \frac{m}{2\pi\hbar t} dz$ (2.12) for the probability density to observe a quantum particle an instant after the moment of its release would change if one replaces an expression for the phase $S(z, z', t)$ in the equation $G(z, z', t) = \sqrt{\frac{m}{2\pi i\hbar t}} \exp\left(iS(z, z', t)/\hbar\right)$ (2.10) by another function.*

We could also conclude from Exercise 2.2 that a complicated expression in the exponent in the wave-function in equation (2.10) seems to be useless here, and it does not enter into the expression for the probability density to physically observe the particle.

In order to clarify all these issues, let us slightly modify the initial distribution of the probability density $\Psi_0(x)$ to observe the particle. Namely, let us assume that the particle is localized in a nonvanishing vicinity σ of a position $z = z'$:

$$\Psi_0(x) = \frac{1}{\sqrt{\sigma\sqrt{\pi}}} \exp\left(-\frac{(z - z')^2}{2\sigma^2}\right). \tag{2.13}$$

A factor $1/\sqrt{\sigma\sqrt{\pi}}$ in the expression (2.13) in front of the exponent is introduced in order to provide that the total probability needed to observe the particle somewhere is unity:

$$\int_{-\infty}^{+\infty} |\Psi_0(x)|^2 dx = 1. \tag{2.14}$$

Now, a reader, advanced in dealing with Gaussian type integrals[4], could perform the following useful exercise.

Exercise 2.3. *Verify that in the case of the initial distribution, given in the equation* $\Psi_0(x) = \frac{1}{\sqrt{\sigma\sqrt{\pi}}} \exp\left(-\frac{(z-z')^2}{2\sigma^2}\right)$ *(2.13), the probability density to observe a particle at a given height* z *and a given time* t *equals:*

$$|\Psi(z,t)|^2 = \frac{1}{\sqrt{\pi}\sigma\sqrt{1+\hbar^2 t^2/(m^2\sigma^4)}} \exp\left(\frac{(z-z'+Mgt^2/(2m))^2}{2\sigma^2(1+\hbar^2 t^2/(m^2\sigma^4))}\right). \tag{2.15}$$

The latter expression (2.15) for the probability density to observe the particle as a function of height and time seems much easier to accept than the expression (2.12), which predicts an infinitely fast spreading of a wavefunction, which is precisely localized at the zero time. Indeed, here a *wavepacket*, initially localized in the vicinity of a height $z = z'$ with an uncertainty σ, changes its localization with time.

Exercise 2.4.   *Derive the law of free fall of the center of the Gaussian-type wave-packet defined in the expression* $|\Psi(z,t)|^2 = \frac{1}{\sqrt{\pi}\sigma\sqrt{1+\hbar^2 t^2/(m^2\sigma^4)}} \exp\left(\frac{(z-z'+Mgt^2/(2m))^2}{2\sigma^2(1+\hbar^2 t^2/(m^2\sigma^4))}\right)$ *(2.15).*

---

[4]Courant, R., and Hilbert, D. (1953). *Methods of mathematical physics 1*, Interscience Publ., New York.

The center of this wave-packet moves along the following line:

$$z = z' - Mgt^2/(2m), \qquad (2.16)$$

that is simply along the *classical free fall* trajectory. Thus the center of the wave-packet falls down according to the classical law, and this observation means good news as it provides a bridge to the well-established and understood world of classical mechanics.

A new point in the quantum treatment is that the width $\sigma(t)$ of the wave-packet *evolves* with time:

$$\sigma(t) = \sigma\sqrt{1 + \hbar^2 t^2/(m^2\sigma^4)}. \qquad (2.17)$$

As follows from equation (2.17) the characteristic time of increasing the wave-packet width by a factor of two compared to its initial width is equal to $t_2 = \sqrt{3}m\sigma/\hbar$. The smaller the initial width, the faster the wave-packet spreads. This statement agrees with our previous findings (but it does not agree at all with classical common sense!) that a precisely localized particle (with the wave-packet width equal zero $\sigma \to 0$) instantly spreads to the size of the Universe.

(*Note*: An attentive reader has probably noticed that this statement contradicts incompatibly to the special theory of relativity, which strictly limits the maximum possible speed of such spreading by the speed of light. This contradiction means that nonrelativistic quantum mechanics, which we are discussing here, has to be significantly upgraded in order to account for relativistic effects. Such a theory, known as relativistic field theory, is incomplete nowadays.)

So the novel effect of a purely quantum nature consists of the spreading of an initially localized wave-packet.

Exercise 2.5. *Calculate the characteristic time needed for increasing twice the width of a neutron wave-packet, provided that its initial localization is: i) 10 nm; ii) 10 μm; iii) 10 mm.*

Fig. 2.1   Free fall of quantum fish.

The kind of fish shown in figure [2.1] are known for one specific property: while danger is approaching they expand. The closer they are to the enemy, the larger their size. Probably, they are trying to fear the enemy in this funny way.

Similarly, the spreading of an initially localized wave-packet increases as a function of the fall time.

Let us analyze what makes a wave-packet center move according to the classical law, and also what makes a width of a wave-packet spread. According to equation (2.9), the wave-function at a given height z could be thought of as a *sum of contributions* of point-like sources, located at different heights z' and having the "intensity" of the initial wave-function

$\Psi_0(z')$ at the corresponding height z'.

Contributions of these different point-like sources to the final wave-function amplitude at a height z are proportional to the product of the amplitude of the zero-time wave-function $\Psi_0(z')$ and the *propagator function* $G(z, z', t)$.

The *phase* of the propagator $G(z, z', t)$ changes rapidly from one height z' to another height. This rapid change imposes qualitatively that such a summation $\int G(z, z', t)\Psi_0(z')dz'$ of the probability amplitudes *to propagate from* a height z' to a height z during time t yields a vanishing result almost everywhere, except for those areas where the change of phase is slow.

(*Note*: A mathematically rigorous method to calculate such an integral of a rapidly oscillating function is given in the so-called *stationary phase method*[5], which consists of finding a few points [or areas], which provide dominant contributions to the integral.)

Let us find the area, which yields the main contribution to the integral (2.9). The phase always changes "slowly" in the vicinity of its extremum, thus we should calculate the phase of the expression $G(z, z', t)\Psi_0(z') \sim \exp(iS(z, z', t))$ and find its extremum:

$$\frac{dS(z, z', t)}{dz'} = 0. \tag{2.18}$$

It follows from equation (2.18) that

$$(z - z' + Mgt^2/(2m)) = 0. \tag{2.19}$$

This condition (2.19) means that the extremum of a phase $S(z, z', t)$ is achieved along the classical free-fall trajectory $z = z' - Mgt^2/(2m)$, and this trajectory provides the dominant contribution to the resulting integral in the expression (2.9). Thus the *interference* of different probability amplitudes $G(z, z', t)\Psi_0(z')$ *to propagate* from a height z' to a height z during time t results in the vanishing of contributions of all other trajectories of the center of a wave-packet, except for the classical one.

One could recognize here the least action principle.

---

[5]Courant, R., and Hilbert, D. (1953). *Methods of mathematical physics 1*, Interscience Publ., New York.

This interpretation outlines two important issues.

The first issue is that classical motion of a particle, as we see it, is in fact a result of *interference* of waves, produced by multiple point-like sources. A result of this interference is that the probability to observe the particle, as a function of coordinates and time, is localized along the classical trajectory. This interpretation came from Richard Feynman[6] and it is indeed fruitful: it provides an *explanation* why classical mechanics obey this very elegant principle of least action.

We propose a useful exercise to the reader:

Problem 2.3. *Verify that the phase $S(z, z', t)$ in the expression $G(z, z', t) = \sqrt{\frac{m}{2\pi i \hbar t}} \exp\left(iS(z, z', t)/\hbar\right)$ (2.10) is a classical action.*

Thus, classical motion itself could be thought of as a kind of illusion, hiding the phenomenon of quantum interference.

The second issue is that the above-mentioned localization of the probability density to observe the particle along the classical trajectory takes place only approximately.

There is always a *certain area* around the classical trajectory in the space-time manifold, where the probability to observe the particle is not negligible. The width of such a "strip", where the signs of different contributions are equal to each other and which thus contribute coherently to the integral, is given in the following expression, according to the stationary phase method:

$$\sigma(t) = 1/\sqrt{|d^2\varphi/dz'^2|}, \qquad (2.20)$$

$$\varphi(z, z', t) = S(z, z', t) + i\frac{(z - z')^2}{2\sigma^2}. \qquad (2.21)$$

The sense of a variable $\varphi(z, z', t)$ in the expression (2.21) is that it stands for the phase of the product $G(z, z', t)\Psi_0(z')$. One could verify that the resulting expression for the width $\sigma(t)$ coincides exactly with the

---

[6]Feynman, R. P. (1948). Space-time approach to non-relativistic quantum mechanics, *Rev. Mod. Phys.* **20**, p. 367.

expression given in the equation (2.17). It includes a new effect, which is not described in classical theory.

One could see that the effect of *spreading* is described in a term, proportional to the square of the reduced Planck constant $\hbar^2$ in the equation (2.17). The corresponding rate of spreading is given in the expression:

$$1/\tau^2 = \frac{\hbar^2}{m^2\sigma^4},\qquad(2.22)$$

where τ is a characteristic time of spreading.

The fact that this term (2.22) is proportional to the square of the Planck constant $\hbar^2$ indicates that it is of purely quantum nature. The smaller is the *inertial* mass m, the faster is the spreading. One could also notice that a narrower initial distribution σ results in a faster spreading. In the limit of zero initial localization $\sigma \to 0$ the spreading rate is infinitely large.

This observation agrees with our previous finding (2.12) that in the case of the absolute localization of the height of the particle at time t_0, this particle could be observed with an equal probability at any height after whatever short period of time.

Now we are well prepared to perform a Galileo-type experiment with a quantum particle. Indeed, the expression (2.15) describes the problem of quantum fall of an initially localized wave-packet, which stands for the quantum particle. Such an initially localized quantum particle falls down in the gravitational field of the Earth and reaches a bottom surface after some time delay t_1. While detecting the particle, could we say that the time of fall is equal to t_1?

No!

Indeed, if we repeat this experiment again with exactly the same initial conditions, the measured time of fall would have a *different value* t_2. After many repetitions of this experiment, we would find that the time-of-fall values are distributed around a mean value $\bar{t}$ and their dispersion has a certain width. The mean value is equal to the classical time of fall $\bar{t} = \sqrt{2z_0/g}$. By means of increasing the number of measurements N we could estimate the mean value $\bar{t}$ with any accuracy desired.

(*Note*: The accuracy of measuring the mean value of a random variable, which is thus scattered according to the Gaussian distribution, is inversely proportional to the square root of the number of identical measurements.)

Thus the quantum mechanical approach does not provide a defined result of a single experiment; it gives us rather the probability distribution of results of *numerous* identical experiments.

There is nothing hidden in the quantum equation of motion, in the Schrödinger equation, that would hint to us how to relate a wave-function and the results of a physical experiment. In order to define such a correspondence, one has to supply an additional principle, which would introduce the *probabilistic interpretation* of the wave-function and which relates its modulus square with a corresponding probability density to observe the *particle*.

The only trace of quantum phenomena in the "leaning tower" quantum experiment is the *quantum dispersion* of the times of free fall. Note, however, that one should also expect to observe a dispersion of measured values of the time of free fall in a classical experiment. What is the difference between these two cases?

Classical dispersion arises due to the stochastic uncertainty in the initial height and vertical velocity of a classical object, in uncertainties of measurements, as well as in the external perturbations during the fall. With a properly designed and performed experiment you could in principle decrease each such uncertainty separately, and you could also decrease all these uncertainties simultaneously.

Could we reasonably assume that the quantum dispersion, or the quantum spreading, is nothing else but another manifestation of analogous stochastic uncertainties related to the finite accuracy of any of our experimental procedure or knowledge? No, this hypothesis is not correct; there is a *principal difference* between classical and quantum uncertainties.

This difference is clearly seen in the equation (2.17) because an attempt to decrease the uncertainty of the initial height of a quantum particle would not result in a reduction of the uncertainty of the free fall times! In contrast, it would result in a rapid enhancement of the wave-packet spreading, and thus an enlarging of the dispersion of detected values.

It is *principally impossible* to achieve any small value of the uncertainty in the initial height and simultaneously any small uncertainty in the time of fall distribution for a given initial height z_0. In an extreme case of a very small uncertainty in the initial height, the spreading of the wave-packet during the fall time could become even larger than the initial height z_0.

In order to clarify this issue, we propose the reader to solve the following problems.

Problem 2.4. *You have at your disposal a "leaning tower" with a height H, at the top of which you can shape a wave-packet in the following form* $\Psi_0(x) = \frac{1}{\sqrt{\sigma\sqrt{\pi}}} \exp\left(-\frac{(z-z')^2}{2\sigma^2}\right)$ *(2.13) with the initial width equal σ. You drop this wave-packet from the top of the tower and detect the time intervals of free fall.*

Calculate the distribution of values of the time of fall.

Problem 2.5. *Find the width of an initially shaped wave-packet, which would minimize the spread of the above-mentioned time distribution for a given height of fall H.*

Problem 2.6. *What would be the accuracy of estimating the ratio M/m of the gravitational mass to the inertial mass, if you repeat such an experiment N times?*

Note: Assume that all other uncertainties except for quantum ones are negligible.

Let us look at the problem of quantum spreading of a wave-packet from another point of view.

We suggest a classical analogy to a wave-packet spreading, namely we propose to imagine an explosion of a firework charge in the sky. A cloud of particles with uniformly distributed values and directions of velocities produces a beautiful shining sphere while spreading. In order to reproduce a quantum effect of spreading, one should assume that the maximum velocity of a firework particle is inversely proportional to the initial spatial size of the firework charge.

The extent to which this analogy could be used is limited, but it provides us with a hint to study the *velocity distribution* associated with a wave-packet.

First, let us analyze what we could say about the momentum of a wave-packet. We have already mentioned that the wave-function $\Psi_{\vec{p}}(\vec{r}, t) = \exp(-iE_p t/\hbar + i\vec{p}\vec{r}/\hbar)$ corresponds to a state, in which the momentum $\vec{p}$ of a particle is precisely defined. Such a wave-function is called a plane-wave. We have already mentioned that the corresponding momentum $\vec{p}$ of a particle could be found by means of acting on such a plane-wave function by a *momentum operator* $-i\hbar\vec{\nabla}$:

$$-i\hbar\vec{\nabla}\Psi_{\vec{p}}(\vec{r},t) = \vec{p}\Psi_{\vec{p}}(\vec{r},t). \qquad (2.23)$$

It is said that the function $\Psi_{\vec{p}}(\vec{r},t)$ is an eigenfunction of a momentum operator, while $\vec{p}$ is an eigenvalue of a momentum operator.

Let us try to construct a given wave-packet as a superposition of waves $\Psi_{\vec{p}}$ (2.3) with different momenta $\vec{p}$. An important Fourier theorem tells us how to construct a given wave-packet from plane waves

$$\Psi_0(\vec{r}) = \frac{1}{(2\pi)^{3/2}} \int C(\vec{p})\exp(i\vec{p}\vec{r})d^3p, \qquad (2.24)$$

with coefficients $C(\vec{p})$ equal:

$$C(\vec{p}) = \frac{1}{(2\pi)^{3/2}} \int \Psi_0(\vec{r})\exp(-i\vec{p}\vec{r})d^3r. \qquad (2.25)$$

Exercise 2.6. *Verify that the following equality is valid:* $\int |C(\vec{p})|^2 d^3p = \int |\Psi_0(\vec{r})|^2 d^3r.$

What is a physical sense of the coefficient $C(\vec{p})$?

An answer to this question follows from the probabilistic interpretation of the modulus square of a wave-function (an interested reader could try to motivate it). It consists of the statement that the modulus square of this coefficient $|C(p)|^2$ stands for the probability to observe a *momentum value* $\vec{p}$ in a state Ψ_0. We see that the coefficient $C(\vec{p})$ characterizes the same state Ψ_0, but it calculates a momentum distribution instead of calculating a spatial distribution.

A momentum distribution is an equivalent description of a quantum state; it is said that the coefficient $C(p)$ gives a wave-function *in the momentum representation*.

Exercise 2.7. *Find the momentum distribution in a Gaussian-type wave-packet* $\Psi_0(x) = \frac{1}{\sqrt{\sigma\sqrt{\pi}}} \exp\left(-\frac{(z-z')^2}{2\sigma^2}\right)$ *(2.13).*

We write down the solution:

$$C_0(p) = \sqrt{\frac{\sigma}{\pi\hbar}} \exp\left(-\frac{p^2\sigma^2}{2\hbar^2}\right). \tag{2.26}$$

The width of the momentum distribution is:

$$\Delta p = \hbar/\sigma. \tag{2.27}$$

The above expression (2.27) is very similar to the fundamental expression (2.1). It shows that a wave-packet with the spatial width σ could be thought of as a superposition of plane waves. Each plane wave contributes to the wave-packet with a certain weight. The width of the distribution of these weights is $\hbar/\sigma$.

A physical meaning of the expression (2.26) is straightforward: the smaller the spatial width of a wave-packet, the larger its spreading in the momentum space. This momentum dispersion of a wave-packet "explains" the spreading of an initially localized wave-packet. Indeed, the corresponding *velocity* distribution width is equal $\Delta v = \Delta p/m = \hbar/(m\sigma)$. Thus the rate of spreading is inversely proportional to the spatial size σ and the *inertial* mass m.

We have just investigated a particular case of a more general and famous statement called an *uncertainty relation*. The uncertainty relation states that the product of the spatial dispersion and the momentum dispersion could not be smaller than a half of the reduced Planck constant $\hbar/2$. This relation is a consequence of the wave nature of quantum objects; it states that a momentum and a coordinate of a particle are not completely independent variables.

Moreover, the quantum mechanical description allows either spatial variables *or* momentum variables to be a mean of description of a physical state.

Let us think how to find a mean value of the momentum in a given state $\Psi_0(x)$.

The answer is simple: One should construct a corresponding momentum representation function $C(\vec{p})$ and calculate a mean value, according to the general rule of finding a mean value of some random variable, provided we have a probability density for this variable:

$$< \vec{p} >= \int \vec{p} |C(p)|^2 d^3 p. \tag{2.28}$$

As we decided to study the momentum-side of the story, let us find out what happens with the momentum of a quantum particle in a gravitational field.

Problem 2.7. *Verify that in the momentum representation the propagator $G(z, z', t) = \sqrt{\frac{m}{2\pi i \hbar t}} \exp\left(iS(z, z', t)/\hbar\right)$ (2.10) looks as follows:*

$$G(p, p', t) = \exp\left(-i\frac{t}{2m\hbar}(p^2 - pMgt + 1/3M^2g^2t^2)\right)\delta(p - p' + Mgt). \tag{2.29}$$

One could see that the momentum distribution of an initially shaped wave-packet "is shifted" during quantum fall according to the classical law by the value equal Mgt.

Problem 2.8. *Calculate the kinetic energy distribution in a state given in the equation $\Psi_0(x) = \frac{1}{\sqrt{\sigma\sqrt{\pi}}} \exp\left(-\frac{(z-z')^2}{2\sigma^2}\right)$ (2.13), and analyze what happens to the kinetic energy distribution during the quantum fall of an initially shaped wave-packet.*

Concluding our first finding about a quantum fall experiment, we outline that the difference between quantum and classical free falls is not that dramatic, in spite of the dramatic difference between classical and quantum concepts. Namely, the center of a wave-packet moves along the classical trajectory, and this fact establishes a relation between quantum and classical mechanics. As soon as a wave-packet size could be neglected, we return to the classical picture.

Perhaps, one would not be satisfied with the fact that the only quantum effect in the case of free fall consists of the spreading of a wave-packet, and therefore one would like to see more spectacular evidence of the wave nature of quantum objects. In the following we are going to see how a quantum particle, bouncing on a surface, could help us in exposing such quantum properties. In particular, we are going to study in more detail the central quantum mechanical concept, the concept of a wave-function.

2.2 Persistence of States

The difference in the characteristic sizes associated with three typical examples (atoms, apples and planets) mentioned repeatedly in the previous sections is enormous. However, an attempt to apply our knowledge, which we have gained while dealing with apples, to atoms on one hand and to planets on the other hand is a natural and often fruitful approach in science. Thus, in the case of planets and apples this approach resulted in the development of Newtonian mechanics.

Nevertheless, an apparently reasonable guess that atoms, like apples or planets, should follow the same laws of classical Newtonian mechanics, turned out to be an illusion.

In the early twentieth century the striking discrepancy between classical expectations and experimental facts concerning lifetimes of constituents of the atom was discovered. One of the first historical tasks of quantum mechanics was to explain the *stability of atoms*.

According to classical electrodynamics and Newtonian mechanics, an electron should promptly fall on the nucleus due to emitting electromagnetic waves. This fall is possible due to the fact that a classical electron could have any large negative energy in the attractive field of the nucleus. Thus it would continue emitting electromagnetic waves that would continue bringing the energy away. Consequently, the electron would approach closer and closer to the nucleus.

Classical estimations show that the time of life of an electron in an atom is small on a microscopic scale. This statement is in obvious contradiction with the observed stability of our world.

Quantum mechanics propose a beautiful solution for this problem. Briefly, it consists of the statement that an electron could have only well-defined minimum energy in the atom, and in this *state* it cannot emit energy anymore.

This quantum explanation could be understood qualitatively in terms of the *uncertainty relation* between a coordinate and a momentum of a quantum particle.

Indeed, the fact of spatial localization of an electron in a small vicinity σ around the atomic nucleus means large (by its absolute value) negative potential energy. However, *spatial* localization, according to the uncertainty relation, results in a spread in the *momentum* distribution with a characteristic width $\hbar/\sigma$, and thus with a corresponding spread in the kinetic energy equal $\hbar^2/\sigma^2$. This spread of (positive) kinetic energy compensates the negative potential energy in such a way that there is a minimum in the total energy of the electron.

However, these qualitative arguments, though they might be of interest, could not become a convincing reason to accept such a strange theory as quantum mechanics. A crucial role in the triumph of quantum theory was played by a peculiar prediction that the energy of an electron, bound inside an atom, could have only *certain discrete values*. This prediction explained precisely these sharp lines, which had been observed in the atomic spectra and which had no classical explanation.

Before approaching these peculiar features of the energy of a quantum particle, let us question how a wave-function relates to the value of any physical quantity, which characterizes the particle.

We have already studied an example of a particle, which falls down in a gravitational field from a given height, and found that the wave-function of a corresponding state could only provide a probability distribution of the times of fall, or a distribution of momenta of a particle, or a distribution of coordinate values.

However, we know that there are physical values that stay unchanged during the motion of a particle. These conserved physical values play a special role, and they are called the *integrals of motion*. An energy of a closed system is one such integral of motion.

Does the energy conservation contradict the probabilistic outcome of a quantum experiment?

And what does a wave-function looks like in the case of fixed energy of the particle?

The answer to these questions is straightforward: On one hand, quantum states, which are characterized by a given energy, should be described using such a wave-function that it yields the zero probability to observe any other value of the energy of the particle. On the other hand, the probability to observe the given value of the energy of the particle in such a state should be equal unity.

We have already learnt the algorithm of finding a wave-function, which corresponds to a given value of *momentum*. Let us remind and then extrapolate the main steps used.

The corresponding wave-function is an eigenfunction of a *momentum operator* $\hat{p} = -i\hbar\vec{\nabla}$, which means that it is a solution of the following equation:

$$-i\hbar\vec{\nabla}\Psi_p(\vec{r}) = \vec{p}\Psi_p(\vec{r}), \tag{2.30}$$

while $\vec{p}$ is called an eigenvalue of the momentum operator $-i\hbar\vec{\nabla}$. Such a wave-function has the following form:

$$\Psi_p(\vec{r}) = \left(\frac{1}{2\pi}\right)^{2/3} \exp\left(i\vec{p}\vec{r}/\hbar\right). \tag{2.31}$$

We have established that the probability to observe a momentum of a quantum particle in a small "volume" d^3p around a given value $\vec{p}$ in a given state Ψ_0 is determined in the expression:

$$dW_p = d^3p \left| \int \Psi_0(\vec{r})\Psi_p^*(\vec{r})d^3r \right|^2. \tag{2.32}$$

If the given equality holds for the wave-function $\Psi_0 = \Psi_{p_0}$ then the probability density for the momentum value is equal $\delta(\vec{p} - \vec{p_0})$. This statement simply means that the probability to observe any other value of momentum, except for the value $\vec{p_0}$, is equal zero if a particle is settled in the state Ψ_{p_0}.

Now we would like to generalize this approach of calculating the probability distribution of momentum values to the calculation of *any quantum mechanical quantity*, as long as it could be "constructed" from the momentum and coordinates of the particle.

Among other physical quantities we will be interested in energy. *An operator*, which corresponds to the *kinetic energy* should be constructed in such a way that the classical relation between the kinetic energy and the momentum is kept valid $\widehat{T} = (\hat{p})^2/(2m)$.

Obviously, in the case of motion along one dimension, the kinetic energy operator has the following explicit form: $\widehat{T} = -\hbar^2/(2m)d^2/dz^2$. It is reasonable to conclude that the operator of energy in the case of an external potential $U(z)$ should have the form:

$$\widehat{H} = -\frac{\hbar^2}{2m}\frac{d^2}{dz^2} + U(z). \tag{2.33}$$

This operator is known as *Hamiltonian*; it plays a key role in quantum mechanics.

The Schrödinger equation could be rewritten then as follows:

$$-i\hbar\frac{\partial}{\partial t}\Psi(x,t) = \widehat{H}\Psi(x,t). \tag{2.34}$$

The states with a given energy are eigenfunctions of Hamiltonian:

$$\widehat{H}\Psi_E(x,t) = E\Psi_E(x,t). \tag{2.35}$$

The peculiar feature of such states follows immediately from the equation (2.34).

Exercise 2.8. *Verify that the following equation is valid:*
$$\Psi_E(x,t) = \varphi_E(x)\exp\left(-iEt/\hbar\right). \tag{2.36}$$

Exercise 2.9. *Verify that an energy value E, measured in a given state $\Psi_E(x,t) = \varphi_\lambda(x) \exp(-iEt)$, does not evolve.*

The spatial probability density $|\Psi(x,t)|^2 = |\varphi(x)|^2$ to observe a particle in such a particular state with a fixed energy does not evolve. Such states are called *"stationary states"*, they are also called *"standing waves"*. A peculiar feature of such a state is that it has nothing common with classical motion. Well, we have just used the word "motion"... If the probability density to observe a particle does not evolve in time, is there any motion at all?

Indeed, we found in the previous chapter that a quantum "reason" for motion consists of the spatial and temporal changing of the phase of the wave-function. In the case of a stationary state, the *phase* of the corresponding wave-function is *the same at all heights*. Thus, for a state, which is characterized by a given fixed value of energy, we calculate a stationary distribution of the probability density instead of calculating a wave-packet, which is moving in space and time.

Therefore, it is reasonable to assume that the phenomenon of stationary states manifests clearly pronounced quantum features, and thus one could point a dramatic difference between classical and quantum descriptions of the behavior of particles in such states. An important feature of stationary states is that they are perpetual; they remain unchanged as long as the system of interest is closed (i.e. as long as it is isolated from the external interactions).

In particular, the existence of such stationary states explains the intriguing stability of atoms. While previously studying the evolution of wave-packets, we assumed that the initial wave-function is characterized by a certain form (a gaussian wave-packet, for instance). However, we have not discussed how one could "shape" such a wave-packet. A wave-function in a stationary state provides us an example of a wave-packet, which is naturally settled in a corresponding physical system.

Now we are well prepared to answer in practical terms the question on how to calculate such stationary quantum states. For this purpose we would like to consider again our favourite example of a quantum particle bouncing in a gravitational field above an ideally reflecting horizontal material surface. The Schrödinger equation for a stationary state wave-function (its time-independent part) with a given energy in such a system turns out to be:

$$\left(-\frac{\hbar^2}{2m}\frac{d^2}{dz^2} + Mgz\right)\varphi(z) = E\varphi(z). \tag{2.37}$$

Prior to solving this equation (2.37) we would like to introduce the characteristic *spatial and energy scales* in the above equation. We are going to choose these scales in such a way that all coefficients in the equation (2.37) turn into a unit if being expressed in new dimensionless variables, which are normalized to the introduced spatial and temporary scales. The new dimensionless variables are $x = z/l_0$ and $\lambda = E/\varepsilon_0$, while l_0 and ε_0 are given in the following expressions:

$$l_0 = \sqrt[3]{\frac{\hbar^2}{2mMg}}, \tag{2.38}$$

$$\varepsilon_0 = \sqrt[3]{\frac{\hbar^2 M^2 g^2}{2m}}. \tag{2.39}$$

In order to get some feeling on typical values of these characteristic scales in the cases of particular quantum objects, we propose that the reader performs the following simple exercise.

Exercise 2.10. *Calculate the characteristic gravitational spatial and energy scales $l_0 = \sqrt[3]{\frac{\hbar^2}{2mMg}}$ (2.38) and $\varepsilon_0 = \sqrt[3]{\frac{\hbar^2 M^2 g^2}{2m}}$ (2.39) for the following particles in the gravitational field of the Earth:*
   i) a neutron
   ii) a fulleren molecule C60
   iii) a diamond nanoparticle (approximated as a single-crystal diamond sphere with the diameter of 5 nm, and with the standard bulk diamond density).

In new dimensionless variables, equation (2.37) looks as follows:

$$\left(-\frac{d^2}{dx^2} + x - \lambda\right)\varphi_\lambda(x) = 0. \tag{2.40}$$

The general solution of this equation (2.40) is expressed in a linear combination of two famous special functions, which are known as *Airy functions*[7]:

$$\varphi(x) = C_1 \operatorname{Ai}(x - \lambda) + C_2 \operatorname{Bi}(x - \lambda). \tag{2.41}$$

Note: These two special functions are two independent solutions of the second-order differential equation (2.37). An important property of these functions, to be underlined, is their asymptotic behavior at large positive arguments $x \gg 1$:

$$\operatorname{Ai}(x) \approx \frac{1}{\sqrt[4]{x - \lambda}} \exp\left(-2/3(x - \lambda)^{3/2}\right) \tag{2.42}$$

and

$$\operatorname{Bi}(x) \approx \frac{1}{\sqrt[4]{x - \lambda}} \exp\left(2/3(x - \lambda)^{3/2}\right). \tag{2.43}$$

At the moment we still have too much freedom, which does not allow us to choose the wave-function unambiguously.

However, there are a few more conditions that our wave-function should meet.

First, the probability to observe a particle, bouncing with a given energy above a reflecting surface, in some reasonably large but limited volume should be unit. This condition reads:

$$P = \int_0^\infty |\varphi_\lambda(x)|^2 dx = 1. \tag{2.44}$$

In order to provide the existence of a finite integral in the equation (2.44) the wave-function $\varphi_\lambda(x)$ should *decay* at large x values (it is said

---

[7] Abramowitz, M., and Stegun, I.A. (1972). *Handbook of Mathematical Functions with Formulas, Graphs, and Mathematical Tables*, Dover Publ., New York.

that it should be a square integrable function). This condition imposes that the constant C_2 in the equation (2.41) should be precisely zero ($C_2 = 0$), because $\varphi_\lambda(x)$ does not decay at large x otherwise. Thus the probabilistic interpretation of a wave-function enters into the game and allows us to choose a certain type of solution among many possibilities.

Let us mention that the *modulus* of the constant C_1 in the equation (2.41) could be found also from the equation (2.44):

$$|C_1| = \left(\int_0^\infty |\operatorname{Ai}(x - \lambda)|^2 dx \right)^{-1/2}. \tag{2.45}$$

Second, we have to explicitly take into account the effect of interaction of a particle with a surface. (*Note*: There is not yet any sign of such an effect in the above equations. Let us mention also that the same approach was usually chosen in the classical case; we only assumed that the interaction of a ball with a surface is brief and elastic without introducing and analyzing any details of such an interaction, and thus it gives only a small correction.)

Here we are going to take into account the particle–surface interaction in the following simplified way.

We assume that the surface is an ideal mirror for the particle and that the particle *does not penetrate inside* the mirror. (*Note*: In formal terms, this assumption is simply equivalent to the statement that the potential energy of the interaction of the particle with the surface is a steep function of the distance between them, and also that the absolute value of this potential energy is much larger than the energy of the particle in a gravitational field.)

In other words, the probability of observing a particle below the mirror surface (inside the mirror) is zero, or $\varphi_\lambda(x \leq 0) = 0$. As far as we are interested in calculating the wave-function *above* the surface, it is enough to require that:

$$\varphi_\lambda(0) = 0. \tag{2.46}$$

As far as the constant C_2 in the equation (2.41) equals zero, the above *boundary* condition (2.46) is equivalent to the following equality:

$$\operatorname{Ai}(-\lambda_n) = 0. \tag{2.47}$$

This is a remarkable equation!

It imposes that the (normalized) energy values λ should meet a particular condition, which is given explicitly in equation (2.47). Then it is reasonable to assume that the values λ of physical energy, being the solutions of this equation, would no longer be some arbitrary positive values (as they really were in the case of classical equations of motion). Instead they should form a set of *discrete* values.

Explicit solution of the equation (2.47) fully confirms the above expectation. The first three roots of this equation are equal to $\lambda_1 = 2.338$, $\lambda_2 = 4.088$, and $\lambda_3 = 5.521$. They demonstrate those properties of energy levels, which were used in the early years of quantum mechanics in order to explain the stability of atoms as well as the properties of their emission spectra. In particular, we note the existence of the lowest energy of a bound system and the discrete set of energy values.

As one could see, these properties are the direct consequence of i) the *integrability condition* (2.44), which allows us to calculate the constants C_1 and C_2, as well as of ii) the *boundary condition* (2.46), which could be met only for special values of λ, if any.

Exercise 2.11. *Find the first three eigenvalues of energy for a neutron, bouncing on an ideal mirror in the gravitational field of the Earth.*

Express these three eigenvalues both in dimensionless energies (i.e. normalized to the characteristic gravitational energy scale ε_0) and in electron-volts (eV).

The corresponding wave-function is:

$$\varphi(x) = C_1 \, \mathrm{Ai}(x - \lambda_n). \tag{2.48}$$

The constant C_1 is called the normalization constant; its modulus could be found from equation (2.14).

Exercise 2.12. *Verify that* $|C_1| = 1/|\operatorname{Ai}'(-\lambda_n)|$.

Fig. 2.2 Standing quantum camel. Profile.

It is instructive to analyze the probability density to observe a particle settled in such a quantum state, say in a pure state with quantum number 3. The shape of the probability distribution is imprinted in the brick wall inside a frame held by a person shown in figure [2.2]. The probability is calculated along the horizontal axis, and naturally the height is measured along the vertical axis with the zero height corresponding to the bottom of the frame.

There are several characteristic features illustrated in this figure, which we would like to underline.

First, the shape of this probability distribution is characterized by three maxima. The number of maxima is equal to the ordinal number of the quantum state. This is not by chance. This feature follows from general properties of Airy functions (2.42). In mathematical terms, an Airy function with an ordinal number incremented by one can be formally obtained from the Airy function with a nonincremented quantum number by means of such its shift towards positive arguments until the function is equal to zero again, provided the condition (2.47).

Exercise 2.13. *Verify the latter statement that an Airy function with an ordinal number incremented by 1 is obtained by shifting the Airy function with a nonincremented quantum number towards positive arguments until the function is equal to zero again.*

Explain this fact in terms of physical properties of a linear potential.

Second, the amplitude of a maximum in the probability distribution to observe a particle increases as a function of height (as a function of the ordinal number). This tendency is in line with a well-known classical dependence, which assumes that the absolute velocity of a particle decreases as a function of its height in a gravitational field; and at the turning height the velocity is even tending to zero. However, no periodic-like probability distribution is, of course, observed in the classical case.

Third, an extraordinary quantum feature with no classical analogue is the presence itself of the roots in the probability density distribution. A zero value of the probability means physically that a particle in a pure quantum state could never be found at this height (i.e. you could observe a particle a bit higher than this point and a bit lower than this point, but never in the point itself). Our classical intuition prompts that the particle has to pass this height "infinitely fast". Ignore your classical intuition...

Finally, it would be instructive to compare eventual graphical styles of presentation of quantum and classical motion of a bouncing particle in a gravitational field above a mirror, which would naturally follow from the nature of these two kinds of motion. If the classical motion looks like a sequence of photos taken at equal intervals similar to figure [1.1], the quantum motion is rather represented by the heights of maximum probabilities to observe the particle.

For the case of a quantum state number three, such maximum probabilities are imprinted in the brick wall inside a frame held by a person shown in figure [2.2]. In contrast to the classical case, the distance between neighbor heights increases, not decreases, with increasing the height. And also one could not tell any longer that this is a set of consequent positions of a particle. The particle is rather appearing at the heights of maximum probability without a defined consequence.

Thus we found that a quantum particle, which is bound in a gravitational field in the vicinity of a reflecting surface, could possess only discrete energy values. In fact this conclusion is valid for any quantum motion in a limited volume.

However, this is not the only pronounced quantum feature of states with a given energy.

Let us study the spatial density distribution to observe a particle, in a state with a certain energy, which is then proportional to the term $| \mathrm{Ai}(x - \lambda_1)|^2$ (2.48).

A surprising fact here is that the probability density to observe a particle at any given positive height z has a finite nonzero value. Indeed, from the asymptotic form of the Airy function (2.42) we learn that in the case of large heights $x \gg \lambda_n$ (and correspondingly in the case of $z \gg Mgl_0\lambda_n$) the probability to observe the gravitationally bound particle is exponentially small. However, it is nonzero:

$$\mathrm{Ai}(x) \approx \frac{1}{\sqrt[4]{x - \lambda}} \exp\left(-2/3(x - \lambda)^{3/2}\right). \qquad (2.49)$$

A particular feature of this asymptotic could be illustrated, for instance, using figure [2.2].

First, let us note that this asymptotic behavior of the probability density distribution to observe a particle above the turning height is the same for any quantum state. Notice a line in figure [2.2], which consists of oblique strokes on the brick wall in the upper part of the picture inside the frame.

This line corresponds to the maximum raise height for a classical particle with the same energy as the energy of a quantum particle in the third quantum state.

The probability density is not cut at this height; it continues and continues higher than that.

This result is paradoxical!

It seems to clearly contradict the fact that a particle is characterized by a definite value of energy E. Indeed, once we detect a particle at a height z such that its potential energy is larger than the total energy of the particle $Mgz > E$, this observation imposes formally that the kinetic energy of the particle $E - Mgz$ is negative. One has never detected a classical ball, bouncing elastically on a surface, at a height z larger than its initial dropping height!

In classical mechanics a negative value of the kinetic energy would correspond to an imaginary velocity, which seems to be senseless. Thus, the values of a height z, for which the kinetic energy of the particle is negative $E - Mgz < 0$, are classically forbidden. This conclusion of classical mechanics follows from the fact that both the coordinates and the velocities have definite values and are used as independent variables while describing classical motion.

In quantum theory only a half of variables could be used simultaneously; in our case they are either coordinates or momenta.

If we use the coordinate description, then the kinetic energy is a function of coordinates $E - Mgz$. However, it would not longer be a positively defined physical value (like the kinetic energy $mv^2/2$ in classical mechanics is). As far as there is no trajectory in quantum mechanics, there is no contradiction between an imaginary velocity and the classical definition of velocity as dz/dt. However, the mean kinetic energy (which keeps its classical sense) is positive.

(*Note:* A classical virial theorem, which is also valid for mean values in the quantum case, states that $\bar{T} = 1/3E$ in the case of a quantum particle, bouncing in a gravitational field, so the mean kinetic energy is:

$$\overline{T} = -\frac{\hbar^2}{2m} \int_0^\infty \varphi_n(z) \frac{d^2}{dz^2} \varphi_n(z) dz > 0.) \qquad (2.50)$$

Thus we conclude that a particle could be observed with nonzero probability "under a gravitational barrier", i.e. at classically forbidden "too large" heights z such that $E - Mgz < 0$.

The probability of observing a particle "under barrier" decreases rapidly with increasing the height z as follows from the asymptotic expression for the wave-function at large heights z (2.49), but it is never zero. This phenomenon, called a *"tunneling effect"*, is one of the most intriguing and counterintuitive effects, which has a purely quantum nature. Below we are going to discuss a realistic experiment in which the tunneling of neutrons through a gravitational barrier could be detected.

Problem 2.9. *A quantum particle is settled in the lowest energy state in the gravitational field of the Earth above a mirror.*

Calculate the probability to observe it under the gravitational barrier: $z > E/Mg$.

Problem 2.10. *Verify that the equation for a stationary state in a gravitational field looks as follows:*

$$\left(\frac{p^2}{2m} + i\hbar\frac{d}{dp} - E \right) C_E(p) = 0, \tag{2.51}$$

while the quantization condition has the following form:

$$\frac{1}{\sqrt{2\pi\hbar}} \int_{-\infty}^{\infty} C_E(p)dp = 0. \tag{2.52}$$

Until this point we have exposed amazing quantum features of stationary states.

Now let us look at stationary states from a different point of view, and thus let us try to get a more intuitive *explanation* of these quantum properties.

For this purpose we are going to use the already-mentioned relation between the classical least action principle and quantum interference phenomena. Namely, inspired by the result for the propagator in a gravitational field (2.10), we would like to search for a solution of the stationary problem in the following form:

$$\varphi(z) = A(z)exp(\pm iS_0(z)/\hbar), \qquad (2.53)$$

$$S(z) = \int_0^z \sqrt{2m(E - Mgz')}dz'. \qquad (2.54)$$

In the above expression (2.53, 2.54), one could recognize the radial part of the classical action (1.28) in the phase $S(z)$. A function $A(z)$ is required there in order to compensate for the difference between an exact quantum solution and the approximate exponent $exp(\pm iS/\hbar)$ function. Substituting the expression (2.53, 2.54) into the stationary Schrödinger equation (2.37) and *neglecting higher-order terms* over the reduced Planck constant $\hbar$ we get the following approximation for the wave-function:

$$\varphi(z) = A(z)exp(\pm iS_0(z)/\hbar), \qquad (2.55)$$

$$S(z) = \frac{2\sqrt{2m} \left[(E - Mgz)^{3/2} - E^{3/2}\right]}{3Mg}, \qquad (2.56)$$

$$A(z) = \frac{1}{\sqrt[4]{2m(E - Mgz)}}. \qquad (2.57)$$

The above approximation (2.55, 2.56, 2.57) is known as the *semiclassical*, or WKB (*Wentzel–Kramers–Brillouin*) approximation[8-10].

This approximation plays an important role in establishing a relation between classical and quantum worlds (this approximation is particularly precise in the case of linear quantum potentials, like that describing a constant gravitational field). This approximation states that wave properties of a quantum particle are described to a high extent in an oscillating imaginary exponent with the phase, which is just equal to the classical action divided by the Planck constant.

In the case of large values of the phase, the exponent changes rapidly while the coordinate passes from one height z to another height. In the

[8]Wentzel, G. (1926). Eine Verallgemeinerung der Quantenbedingungen fur die Zwecke der Wellenmechanik, *Zeit. Phys.* **38**, p. 518.

[9]Kramers, H. A. (1926). Wellenmechanik und halbzahlige Quantisierung, *Zeit. Phys.* **39**, p. 828.

[10]Brillouin, L. (1926). La mechanique ondulatoire de Schrodinger: une methode generale de resolution par approximations successives, *Compt. Rend.* **183**, p. 24.

case of a propagator, this property provides that the center of a wave-packet moves along the classical trajectory as a result of interference of partial waves, which constitute the wave-packet. In other words, one should expect that the semiclassical approximation works well, if the integrand in the phase S is large.

In the opposite case, namely near the heights, at which the condition $\sqrt{E - Mgz} = 0$ holds (they are called the *turning heights [points]* because a classical particle stops or turns at these heights) the integrand is small; and the function $A(z)$ (2.57) starts to grow rapidly in order to compensate for the difference between semiclassical and exact solutions. This feature means that the semiclassical approximation fails near these turning heights.

Thus, purely quantum features should be somehow related to these heights.

Let us reveal the role of turning heights for the existence of a stationary solution. We remember that a wave-function should meet the boundary condition at the turning height $z = 0$, namely the condition $\varphi(0) = 0$. Thus we have to find such a combination of two independent solutions given in equation (2.55) and labeled by the signs plus and minus in the exponent, that the wave-function vanishes at the zero height $z = 0$. This is possible if the two waves have opposite signs.

Such a combination of the two solutions for heights $z < E/Mg$ has the following form:

$$\varphi(z) \sim \frac{1}{\sqrt[4]{2m(E - Mgz)}} \sin\left(S(z)/\hbar\right). \qquad (2.58)$$

One could see that the role of the turning height $z = 0$ is to "create" a reflected wave $\exp(iS/\hbar)$, which has an opposite phase compared to the incoming wave $\exp(-iS/\hbar)$.

What behavior should we expect from a wave-function at another turning height $H = E/Mg$?

This turning height corresponds to the classical height of bouncing of a particle with the total energy in a gravitational field equal to E. A rough approximation for the problem of quantum motion of this particle could be obtained using the following simplified arguments. As far as a classical particle cannot raise in a gravitational field higher than the height H, the wave-function (2.58) should probably also become equal zero at this height H.

The latter condition simply requires that the argument of *sin* function should be equal to an integer multiple of π, namely that $S(H) = \pi n$, with n an integer number.

It is important that this condition is met only for certain values of energy such that:

$$\int_0^H \sqrt{2m(E - Mgz')}dz' = \pi n. \tag{2.59}$$

We have just obtained a formula, which is called in physics a *quantization condition* (2.59). In order to derive it, we used a simplified argument that the semiclassical wave-function should vanish at the classical turning height.

We are aware that this our simplification is rather crude. Indeed, our solution (2.58) for the phase of the wave-function fails near the turning height equal $z = E/Mg$, as we have already mentioned. Therefore, strictly speaking, we are not allowed to use this solution everywhere from the height $z = 0$ up to the height $z = H$; nevertheless, we do. One could show that the effect of the turning height consists in fact of the appearance of an additional phase $\gamma = \pi/4$, [11] by which the value $S(H)$ should be shifted:

$$\int_0^H \sqrt{2m(E - Mgz')}dz' = \pi(n - 1/4). \tag{2.60}$$

$$n = 1, 2, 3... \tag{2.61}$$

Problem 2.11. *Verify that the energy levels of a quantum particle, bouncing in a gravitational field above a reflecting surface, are given within the semiclassical (WKB) approximation in the following expression:*

$$E_n = \varepsilon_0 \left(\frac{3\pi}{4}(2n - \frac{1}{2})\right)^{2/3}. \tag{2.62}$$

---

[11]An elegant way to do so is to allow coordinates getting complex values. For explanation of this and many other issues see [Landau, L.D., and Lifshitz, E.M. (1965) *Quantum Mechanics, A Course of Theoretical Physics, Vol. 3*, Pergamon Press, UK].

Problem 2.12. *Estimate the accuracy of calculating the three lowest energy levels, given in this approximate expression, for instance by comparing the semiclassical (WKB) and precise solutions.*

Thus the quantization of energy follows i) from the *interference* of incoming and reflected waves, and ii) from the fact that a wave-function should promptly *decay* inside the classically forbidden region.

The latter issue requires some additional attention. One could verify that at the heights $z > E/Mg$ larger than the turning height, i.e. in the classically forbidden region, our solution is still valid. We are going to show that in the case of the semiclassical quantization condition (2.60) being met, the solution (2.58) for the phase is transformed into a purely *decaying exponent* $\exp(-S(z)/\hbar)$ inside the classically forbidden region of too large heights $z > E/Mg$.

Thus, the semiclassical quantization condition as it is expressed in equation (2.60) is in fact equivalent to the requirement of the absence of a divergent exponent in the solution of the stationary problem in the classically forbidden region. (*Note:* We have already used an analogous condition, while attributing the zero value to the coefficient $C_2 = 0$ in front of a divergent solution in the expression [2.41] for the exact solution of the stationary Schrödinger equation.)

It is surprising that a semiclassical expression could fairly describe the phenomenon of tunneling of a quantum particle through a potential barrier, and could predict a value of an effect, which seems to have a purely quantum nature. The probability of observing a particle inside the gravitational barrier is small, and it decreases rapidly (exponentially) with increasing the height above the turning heights; however, it is nonzero in the semiclassical approximation.

We see that essential features of classical motion are encrypted in the quantum mechanical formalism. *Classical and quantum worlds are deeply interrelated via the least action principle.*

Quantum features are most brightly exposed in *stationary states*. We have already mentioned that the spatial distribution of the probability density in such states with a defined energy does not evolve. However, we also understand that a classical bouncing ball, dropped from the height H, has a well-defined energy equal to $E = MgH$, provided no (minor) dissipation of energy. However, at the same time we could observe indubitable evolution of the ball position in space.

How does the *transition* between a quantum "motionless" state and a classical bouncing occur?

In order to address the process of this transition, let us study *a superposition of states* with different energies. For example, we will treat a wave-function, which is a sum of wave-functions of the two lowest gravitational states, discussed above:

$$\Psi_{12}(z,t) = \frac{1}{\sqrt{2}} \left[\varphi_1(z) \exp\left(-iE_1 t/\hbar\right) + \varphi_2(z) \exp\left(-iE_2 t/\hbar\right) \right]. \quad (2.63)$$

The modulus square of this wave-function $|\Psi_{12}(z,t)|^2$ is:

$$|\Psi_{12}(z,t)|^2 = \frac{1}{2} \left[|\varphi_1(z)|^2 + |\varphi_2(z)|^2 + 2\varphi_1(z)\varphi_2(z) \cos\left((E_2 - E_1)t/\hbar\right) \right]. \quad (2.64)$$

As one could conclude from equation (2.64), the probability density in the above example is now a function of time due to the presence of the term $2\varphi_1(z)\varphi_2(z) \cos\left((E_2 - E_1)t/\hbar\right)$.

Problem 2.13. *How evolves the probability to observe a certain energy in the quantum state composed of two quantum states with different energies* $|\Psi_{12}(z,t)|^2 = \frac{1}{2} \left[|\varphi_1(z)|^2 + |\varphi_2(z)|^2 + 2\varphi_1(z)\varphi_2(z) \cos\left((E_2 - E_1)t/\hbar\right) \right]$ *(2.64)?*

Problem 2.14. *Verify that the probability density to observe a particle in the vicinity of a reflecting horizontal surface in the quantum state, described above, is equal to:*

$$|\Psi_{12}(z,t)|^2 \cong \frac{z^2}{l_0^3} \left(1 + \cos\left((E_2 - E_1)t/\hbar\right)\right). \quad (2.65)$$

Thus, an effect of motion in the considered system (motion in quantum mechanics is understood in terms of the evolution of the probability density to observe a particle in a certain point at a certain moment) is related to the interference of states with different energy; a particle does not have a certain energy in the mentioned state. We see that a superposition of only two stationary states results in a harmonic time-variation of the probability density.

Now one could try to guess on how to produce an illusion of a classical bouncing of a quantum particle with the total energy E. In order to do that, one could construct a superposition of many stationary quantum states with energies close to a given energy value E_{n_0}. The difference between energy levels of these states, which comprise the wave-packet, should be much smaller than their mean energy:

$$\Psi = \sum_{k=n1}^{n2} \varphi_k(z) \exp(-iE_k t/\hbar), \tag{2.66}$$

$$E_{n_0} \gg |E_{n2} - E_{n1}|. \tag{2.67}$$

As long as the energies E_n of the stationary states are close to the value of their mean energy E_{n_0}, one could approximate these values as follows:

$$E_n \approx E_{n_0} + \frac{dE}{dn}(n - n_0). \tag{2.68}$$

Using this approximation (2.68), we immediately obtain the following result for the wave-packet:

$$\Psi = \exp\left(-iE_{n_0}t/\hbar\right) \sum_{k=-\Delta n}^{\Delta n} \varphi_{n_0+k}(z) \exp(-i2\pi kt/T_{n_0}), \tag{2.69}$$

$$T_{n_0} = \frac{\hbar}{2\pi dE/dn|_{n_0}}. \tag{2.70}$$

Problem 2.15. *Verify the validity of the above expression $T_{n_0} = \frac{\hbar}{2\pi dE/dn|_{n_0}}$ (2.70).*

Physical sense of the characteristic time T_{n_0} is obvious from the above equation (2.70). Indeed, at a time instant $t = NT$, with N being an arbitrary integer number, the wave-packet returns precisely to its initial configuration. Thus, T_{n_0} is a *period of quantum bouncing* of a particle settled in a superposition of states with quantum numbers n in the interval $n_0 - \Delta n \leq n \leq n_0 + \Delta n$, provided that the energy spreading within the interval is relatively small $\Delta n \ll n_0$.

Problem 2.16. *Verify, using the semiclassical approximation for quantum levels $E_n = \varepsilon_0 \left(\frac{3\pi}{4}(2n - \frac{1}{2})\right)^{2/3}$ (2.62), that the period $T_{n_0} = \frac{\hbar}{2\pi dE/dn|_{n_0}}$ (2.70) obeys the known classical law:*

$$T_{n_0} = 2\sqrt{\frac{2H_{n_0}}{g}}, \qquad (2.71)$$

where $H_{n_0} = E_{n_0}/Mg$.

We have just established that a quantum wave-packet exhibits periodic motion and its period equals to the classical period.

But what could we say about evolution of the *localization* of such a wave-packet in space?

Solving the following problem confirms that a center of the wave-packet follows the classical trajectory.

Problem 2.17. *Verify that the center of the quantum wave-packet $\Psi = \exp\left(-iE_{n_0}t/\hbar\right) \sum_{k=-\Delta n}^{\Delta n} \varphi_{n_0+k}(z) \exp(-i2\pi kt/T_{n_0})$ with $T_{n_0} = \frac{\hbar}{2\pi dE/dn|_{n_0}}$ (2.69), bounces according to the classical law.*

Note: Use the semiclassical approximation both for the radial wavefunction and for the energy levels.

As a general rule, in order to "construct" a wave-packet, which provides a sort of quantum motion that is similar to classical motion, we have to provide a certain property of the energy spectrum of a system of interest. Namely, we want the system to possess a lot of quantum energy levels, so dense that the spread in energies of stationary states would be much smaller (and thus negligible in the classical limit) compared to their mean energy.

We could notice in the properties of the energy spectrum of a linear gravitational potential (2.62) that these levels become more and more dense with increasing the quantum number. The larger the quantum state energy, the smaller the energy gap between neighbor states, and thus the larger the period T_{n_0} according to equation (2.71). This observation states that the bouncing period increases with increasing the bouncing height $T = 2\sqrt{2H/g}$.

Thus, in order to reproduce the effect of classical bouncing, one wants high enough energy E_{n_0} (and thus large enough height H_{n_0} compared to the characteristic quantum scale l_0 in a gravitational field, and correspondingly a large quantum number $n_0 \gg 1$).

In the opposite case of bouncing particles settled in a few lowest gravitational quantum states with energies of the order of the characteristic gravitational energy scale ε_0, no "classical-like" wave-packet could be constructed. If you shape a superposition of a few such lowest quantum states, then the time evolution of the probability density to observe a particle described in such a wave-packet would have nothing in common with classical motion.

However, even in the case of such parameters of the described quantum interference problem, which provide a sort of quantum bouncing that is rather similar to classical bouncing on a surface, the correspondence between the motion of the wave-packet center and the classical trajectory could not last for too long. Indeed, this nearly classical periodicity, which we have got from the approximative expression (2.71), would be *violated over some time*.

This statement becomes clear as far as we take into account the second order terms in the expansion of the energy levels over the quantum state number:

$$E_n \approx E_{n_0} + \frac{dE}{dn}(n - n_0) + \frac{1}{2}\frac{d^2E}{dn^2}(n - n_0)^2. \tag{2.72}$$

At characteristic times, t, such that the following equality is approximately valid, $t \sim \frac{\hbar}{d^2E/dn^2}$, a contribution of the second-order term in the equation (2.72) starts to be large enough in order to set different quantum states involved in the wave-packet out of equal phase. Thus, at this characteristic time scale, the density distribution to observe a quantum particle would have nothing in common with a classical-like localized behavior of the wave-packet.

It is curious, however, that at an even larger time scale equal to $T_r = \frac{4\pi\hbar}{d^2E/dn^2}$, which is called the *revival* time, the system returns again to the classical-like behavior[12]. Indeed, the contribution of the second-order terms in the expression (2.72) to the phase of the wave-packet is equal to the integer multiple of 2π, thus it exhibits a periodic behavior of another nature.

With all these given examples and particular cases, we have just confirmed our previous conclusion that classical motion is a kind of illusion that is produced by the phenomenon of interference of the probability waves.

However, the quantum world provides us with a much broader choice of different phenomena, which manifest themselves in various types of spatial and temporal dependencies of the probability density to observe the particle. The classical type of motion is only one of many such possibilities, only limiting case of the quantum motion. It occurs as far as many quantum states with close values of energy are combined into a superposition, in a wave-packet of a special type.

In the opposite case of one pure stationary state we get a type of motion, for which quantum properties are particularly exposed.

After having discussed peculiar features of quantum stationary states, we could finally clarify the physical sense of spatial (2.38) and energy scales (2.39) involved.

One could say that a gravitational spatial scale l_0 corresponds to the *characteristic length of a wave*, the motion of which is determined by a gravitational field only. This is a purely quantum characteristic of motion in a gravitational field. The quantum side of a particle free fall is characterized by this scale, namely by such values as a characteristic spatial size

---

[12]Robinett, R. W. (2004). Quantum wave packet revivals, *Phys. Rep.* **392**, p. 1.

of the lowest quantum state, a *length of penetration* "under a gravitational barrier", etc.

A gravitational energy scale ε_0 is another *independent* quantum characteristic of motion. It also gives the time scale of a quantum fall via the energy–time relation (2.2):

$$t_0 = \hbar/\varepsilon_0. \tag{2.73}$$

This time scale gives a typical time period of a wave, which is determined by a gravitational field.

Return to the quantum fall of a wave-packet. We remember that the probability density to observe a falling particle is localized around the classical trajectory within a "strip" of a certain width in the space-time manifold. We see that the two scales, l_0 and t_0, give us the typical width of such a strip.

We already found that characteristic gravitational scales (2.38, 2.39), as well as all properties of stationary states in a gravitational field, change as a function of the value of the mass of a bouncing particle, as clear from the given expressions.

Does this observation mean that the equivalence principle is violated in the quantum case?

We are going to discuss these issues in the next section.

2.3 The Principle of Relativity and the Equivalence Principle in Quantum Mechanics

Once acquainted with some peculiar features of quantum motion, we arrived at the conclusion that classical motion along a certain trajectory is a sort of illusion.

Indeed, classical (particle) motion could be deduced at certain conditions from quantum (wave) motion as its limiting case. In particular we argued that the phenomenon of interference of the (probability) waves enhances constructively some waves in the vicinity of the classical trajectory and cancels them out destructively far from the classical trajectory. This wave nature of quantum motion manifests itself in a lot of effects surprising for classical intuition.

One such quantum mechanical feature, with no straightforward classical analogy, consists of the unavoidable spreading of wave-packets. Thus a quantum object, which is represented as a wave-packet, might be localized initially in a limited spatial volume; then it is released from this volume. As soon as the wave-packet is released, it is going to start spreading. Moreover, it will continue spreading in such a way that the initial spatial dispersion will increase and increase.

The opposite extreme case, a pure stationary state, is another quantum mechanical feature. We understood that a spatial distribution of the probability to observe a quantum particle in a stationary state does not evolve at all. Such a state could be settled in a potential well, which does not evolve. In this case a particle is localized in such a potential trap for any time. Moreover, the energy of a particle in such a stationary state is precisely defined.

In spite of a net difference between quantum and classical motions, a close analogy between them could also be achieved, for instance in the case of the motion of a wave-packet comprised of quantum states with close energy values. The center of such a wave-packet is propagating "along the classical trajectory" at least for relatively short observation times. However, its spreading at sufficiently long observation times breaks the similarity with classical motion.

Thus, quantum motion, thought of as the time evolution of the probability density to observe an object, is a much more general concept than classical motion along a trajectory.

At this point we return to *Galileo's principle of relativity*. The reader remembers that the ability to look to the same physical phenomenon from different frames of reference turned out to be a very fruitful approach for studying properties of classical motion. In particular we could add to the motion of any classical object the motion along a straight line at any given constant speed simply by means of observing this object from another, specially chosen frame of reference.

Such a "boost", generated by the change of the point of view, should not result in a change of any real physical properties according to Galileo's principle.

What would happen if we boost a *quantum* object by means of observing it from a moving frame of reference?

In other words, if we have a solution $\Psi_1(z, t)$ of the Schrödinger equation in one inertial reference frame, K_1, how would it look in another inertial reference frame, K_2?

(*Note*: Let us outline that one frame of reference is moving classically relative to another frame of reference, while the motion of a particle relative to the new frame of reference is of quantum nature. Thus, we are looking on a classical boost of a quantum particle.)

Problem 2.18.   *Verify that the Schrödinger equation with a potential term $U(z)$, written in a new inertial reference frame K_2 with coordinates z' and t', which are related to "old" coordinates in a reference frame K_1 via relations $z' = z + vt$ and $t' = t$, looks as follows:*

$$i\hbar\frac{\partial\Psi_1(z',t)}{\partial t} - i\hbar v\frac{\partial\Psi_1(z',t)}{\partial z'} = \left(-\frac{\hbar^2}{2m}\frac{d^2}{dz'^2} + U(z' - vt, t)\right)\Psi_1(z',t).$$

$$(2.74)$$

One could notice that this equation (2.74) does not look like a typical Schrödinger equation, such as equation (2.6). In contrast to all cases considered above, here the term $-i\hbar v\frac{\partial\Psi_1(z',t)}{\partial z'}$ *breaks a formal covariance of* the Schrödinger equation.

This is a surprising fact!

We remember that the principle of relativity is a cornerstone of any physical theory. The fact that a well-established physical law is expressed in an equation, which is not covariant under Galileo's transformations, is of major concern.

We have already learnt another striking example of physical laws, namely Maxwellian equations for the electromagnetic interactions, which turned out to be not covariant under Galileo's transformations. As the reader knows, in order to resolve this electrodynamical paradox, one had to invent a special theory of relativity, which uses a more general type of coordinate-time transformations, namely the Lorentz transformations, that keep *all* physical laws covariant.

Do we need to invent something equally profound in order to resolve the present quantum paradox?

No, the way out of the present difficulty consists of understanding that the Schrödinger equation is an equation for a wave-function, while only the *modulus square* of this wave-function has a physical sense. Thus, even if the Schrödinger equation looks different in another inertial frame of reference, this difference does not automatically mean that a physical phenomenon is modified as a function of the choice of an inertial frame of reference.

In order to provide the physical equivalence of inertial frames of reference, we want the probability of observing a particle in the same place at the same time to be equal in both frames of reference: $|\Psi_2(z' - vt, t)|^2 = |\Psi_1(z, t)|^2$.

This condition means that two wave-functions, which differ from each other by a factor $\exp(iS(z, t))$, are equivalent (here S is a real-valued function). Thus we could try to "restore" the standard form of the Schrödinger equation by means of writing it down for an *equivalent* wave-function with a properly chosen phase $S(z, t)$:

$$\Psi_1(z', t) = \exp(iS(z', t))\Psi_2(z', t). \tag{2.75}$$

Exercise 2.14. *Verify that such a new equation has the following form:*

$$i\hbar \left[\frac{\partial}{\partial t} - \left(v + \frac{\hbar}{m} \frac{\partial S}{\partial z'} \right) \frac{\partial}{\partial z'} \right] \Psi(z', t) = \left(-\frac{\hbar^2}{2m} \frac{d^2}{dz'^2} + U(z' - vt, t) \right)$$

$$\times \Psi(z', t) - \left(i \frac{\hbar^2}{2m} \frac{\partial^2 S}{\partial z'^2} + + \frac{\hbar^2}{2m} (\frac{\partial S}{\partial z'})^2 + \hbar v \frac{\partial S}{\partial z'} + \hbar \frac{\partial S}{\partial t} \right) \Psi(z', t). \tag{2.76}$$

It is time to profit from the fact that physical observables do not change when one modifies only the phase of a wave-function in the quantum equation of motion. We would like to use this freedom of choosing the phase of a wave-function for modifying the form of the Schrödinger equation. Thus, in order to restore the standard form of the Schrödinger equation we could choose such a phase $S(z', t)$ that all "unwanted" terms would cancel each other out, namely:

$$v + \frac{\hbar}{m} \frac{\partial S}{\partial z'} = 0, \tag{2.77}$$

$$i \frac{\hbar^2}{2m} \frac{\partial^2 S}{\partial z'^2} + + \frac{\hbar^2}{2m} (\frac{\partial S}{\partial z'})^2 + \hbar v \frac{\partial S}{\partial z'} + \hbar \frac{\partial S}{\partial t} = 0. \tag{2.78}$$

The readers familiar with the classical *Hamilton–Jacobi*[13] equation for the classical action, would easily recognize that the equations (2.77, 2.78) are equations for the classical action $\hbar S(z', t)$:

$$\hbar S(z', t) = -mvz' + \frac{mv^2}{2}t. \tag{2.79}$$

Thus, the transformation of a wave-function due to the transition from an inertial reference frame K_1 to an inertial reference frame K_2 turns out to be:

$$\Psi_1(z' - vt, t) = \exp\left(-i\frac{mvz'}{\hbar} + i\frac{mv^2t}{2\hbar}\right)\Psi_2(z', t). \tag{2.80}$$

Problem 2.19. *Find an analogue to the transformation* $\Psi_1(z' - vt, t) = \exp\left(-i\frac{mvz'}{\hbar} + i\frac{mv^2t}{2\hbar}\right)\Psi_2(z', t)$ *(2.80) of the wave-function in the momentum representation.*

Problem 2.20. *Verify that in the case of free motion (i.e. if the potential term is equal zero $U(z) = 0$) the following relations would be valid:*

$$\Psi_2(z', t) = \exp\left(i\frac{p'z'}{\hbar} - i\frac{E't}{\hbar}\right), \tag{2.81}$$

$$p' = p + mv, \tag{2.82}$$

$$E' = E + pv + mv^2/2. \tag{2.83}$$

---

[13]Landau, L.D., and Lifshitz, E.M. (1969). *Mechanics, A Course of Theoretical Physics, Vol. 1*, Pergamon Press, UK.

Thus, we could conclude from the above exercises that the boost of a quantum particle generated by a classical motion of one inertial reference frame (K_2) relative to another inertial reference frame (K_1) is "hidden" in the *phase of the complex exponent* $\exp{(iS(z,t))}$, entering as a factor in the wave-function of the particle. This phase $\hbar S(z,t)$ turns out to possess the *meaning of a classical action*, and this conclusion is imposed by the principle of relativity!

The reader probably remembers that an analogous expression for the classical action (and for the Lagrange function) of a freely moving object was derived by means of following the arguments based on the same principle of relativity.

This result gives us a new glance at the principle of relativity in the quantum world. The equivalence of different inertial frames of reference is expressed for the quantum world by means of the fact that wave-functions in these frames of reference differ from each other only by a phase $S(z,t)$; such wave-functions correspond to *equivalent* distributions of the probability density to observe the particle in a given point at a given moment of time.

The velocity of a classical boost is encoded in this phase of a wave-function via the following relation:

$$\frac{\hbar}{m}\frac{\partial S}{\partial z'} = -v. \tag{2.84}$$

Now we have in our hands all ingredients in order to return to the equivalence principle in quantum mechanics. In order to proceed toward that goal, we would like to explore what the motion of a quantum particle looks like, not only in an inertial frame of reference but also in an accelerated frame of reference.

Here we have to perform the same steps that we did in the case of inertial frames of reference. Let $\Psi_1(z,t)$ be a wave-function of a quantum particle in an inertial reference frame K_1. A reference frame K_2 is accelerated relative to the reference frame K_1; their coordinates are related as follows: $z' = z + vt + at^2/2$.

Let us try to "improve" the noncovariant form of the following Schrödinger equation, written in new coordinates, by means of applying the unitary transformation of the wave-function, which we have already used in the case of inertial frames of reference, namely:

$$\Psi_1(z' - vt - at^2/2, t) = \exp\left(iS(z',t)\right)\Psi_2(z',t). \qquad (2.85)$$

Problem 2.21.   *Derive the Schrödinger equation for the function* $\Psi_2(z',t)$ *in the form, which is analogous to the form of the equation* $i\hbar\left[\frac{\partial}{\partial t} - \left(v + \frac{\hbar}{m}\frac{\partial S}{\partial z'}\right)\frac{\partial}{\partial z'}\right]\Psi(z',t) = \left(-\frac{\hbar^2}{2m}\frac{d^2}{dz'^2} + U(z' - vt, t)\right)\Psi(z',t) - \left(i\frac{\hbar^2}{2m}\frac{\partial^2 S}{\partial z'^2} + +\frac{\hbar^2}{2m}(\frac{\partial S}{\partial z'})^2 + \hbar v\frac{\partial S}{\partial z'} + \hbar\frac{\partial S}{\partial t}\right)\Psi(z',t)$ (2.76) *in a uniformly accelerated frame of reference.*

Again, as in the case of a transition from one inertial frame of reference to another, we are going to try here, for a transition to an accelerated frame of reference, to find such a phase $S(z',t)$ that all unwanted terms in the quantum equation of motion cancel each other out.

The first condition of canceling the terms proportional to a factor $\frac{\partial\Psi_2(z',t)}{\partial z'}$ yields an equation for the phase gradient (compare it with equation [2.77]):

$$at + v + \frac{\hbar}{m}\frac{\partial S}{\partial z'} = 0. \qquad (2.86)$$

From equation (2.86) we find that:

$$S(z',t) = -\frac{m(v + at)z'}{\hbar} + f(t), \qquad (2.87)$$

Here $f(t)$ is a function to be found; it depends only on time.

The second condition allows us to find the function $f(t)$ explicitly. It consists of canceling out the terms in the quantum equation of motion, which contain derivatives of the phase. This condition is analogous to equation (2.78). An important difference between these two cases, however, is that the phase $S(z',t)$ contains a term $-maz't/\hbar$, so time derivatives of the phase $S(z',t)$ depend explicitly on a height z' and thus they could not be canceled out by any choice of the function $f(t)$, which depends only on time, not on height.

Problem 2.22. *Verify that the function $f(t)$ in the equation $S(z', t) = -\frac{m(v+at)z'}{\hbar} + f(t)$ (2.87) should satisfy the following equation:*

$$\frac{df}{dt} = \frac{m(v + at)^2}{2\hbar}. \tag{2.88}$$

Problem 2.23. *Verify that the product $\hbar S(z't)$ is equal to the classical action in the case of a uniformly accelerated object:*

$$S(z', t) = -\frac{m(v + at)z'}{\hbar} + \frac{m}{2\hbar}(v^2 + vat + a^2t^2/3)t. \tag{2.89}$$

Finally, we could easily find out that the Schrödinger equation in a uniformly accelerated reference frame K_2 looks as follows:

$$i\hbar\frac{\partial \Psi_2(z', t)}{\partial t} = \left(-\frac{\hbar^2}{2m}\frac{d^2}{dz'^2} + U(z' - vt, t) - maz'\right)\Psi_2(z', t). \tag{2.90}$$

The unavoidable appearance of the term $-maz'$, which describes the effects of inertia in an accelerated frame of reference is important. Thus, as expected, a wave-function of a particle moving in an accelerated frame of reference is described in the Schrödinger equation, which is not equivalent to the equation in any inertial frame of reference; the wave-function of a particle in an accelerated frame of reference differs by the *inertial interaction term* $-maz'$.

Now we are prepared to get inside a falling elevator and to discuss consequences of the equivalence principle $m = M$ for the quantum mechanical motion.

As one could conclude from equation (2.90), the gravitational potential Mgz is totally compensated with the inertial term $-mgz'$, which appears in an accelerated frame of reference in a falling elevator; and thus the corresponding Schrödinger equation becomes equivalent to an equation for a freely moving particle. Thus we could compensate gravity by acceleration of a frame of reference for any quantum particle, independently of the value of its mass.

The latter observation means that gravity is locally equivalent to an acceleration of a frame of reference (*neglecting the effects of gravitational gradients*), exactly as it is in the case of classical mechanics.

The arguments, which we used above in order to investigate how a quantum particle behaves in a falling frame of reference, are similar to those we used for the transition from one inertial frame of reference to another frame. This similarity outlines the key idea of the equivalence principle, which assumes that a motion of a falling object is a generalization of free motion. These motions are manifestations of pure *geometric properties of space-time.*

In quantum theory, this conclusion is expressed via the condition that the wave-function of a falling object in an inertial frame of reference is equivalent to the wave-function of a freely moving object in a properly chosen frame of reference. This equivalence means that these wave-functions differ from each other only by a *unitary factor* $\exp(iS(z,t))$, which does not affect the observable probability density to find the object in a given place at a given time.

It is worth mentioning that the transformations of a type

$$\Psi_1 \rightarrow \exp(iS(z,t)\Psi_2) \qquad (2.91)$$

are known as *local gauge* transformations; they play a very important general role in physics.

Let us return for a moment to the classical equation of a falling object in a gravitational field with the acceleration g:

$$\ddot{z} = \frac{M}{m}g. \qquad (2.92)$$

As long as the equivalence principle is valid, this equation (2.92) could be treated as a universal geometrical relation between space and time (in the sense that it is the same for any particle independent of its mass value).

This relation is as simple as that: The spatial scale l_0 of classical motion is proportional to the square of the temporal scale t_0 of classical motion. One is free to choose either the spatial scale or the temporal scale, but the relation between them is fixed.

Everything is different in the quantum world.

The Schrödinger equation for a falling particle contains both those characteristic gravitational scales (as it is clear from equations [2.38, 2.39]). We reproduce these equations here with a slight modification, which consists of the replacement of the energy gravitational scale ε_0 by the related temporal characteristic gravitational scale t_0, which is more convenient in the present context:

$$l_0 = \sqrt[3]{\frac{\hbar^2}{2mMg}}, \tag{2.93}$$

$$t_0 = \hbar/\varepsilon_0 = \sqrt[3]{\frac{2m\hbar}{M^2 g^2}}. \tag{2.94}$$

Arguing in terms of the formal dimensional analysis of the variables involved in the Schrödinger equation, one could combine two independent values, one with the dimension of length and another one the dimension of time, respectively (l_0, t_0), using the "ingredients" of this quantum equation of motion because the equation includes not only masses and the gravity intensity but also the fundamental Planck constant $\hbar$, which has the dimension of action.

The universal form of the Schrödinger equation in dimensionless variables $\tau = t/t_0$ and $x = z/l_0$ is:

$$i\frac{\partial \tilde{\Psi}(x, \tau)}{\partial \tau} = -\frac{\partial^2 \tilde{\Psi}(x, \tau)}{\partial x^2} + x\tilde{\Psi}(x, \tau). \tag{2.95}$$

A physical solution $\Psi(z, t) = \tilde{\Psi}(xl_0, \tau t_0)$ includes both inertial and gravitational masses of a particle.

Thus, motion of any quantum particle in a gravitational field assumes that the motion is always associated with a *natural clock* with the characteristic rate unit t_0 and with a *natural ruler* with a characteristic length unit l_0, which are defined by the values of gravitational and inertial masses of the particle, as well as by the gravity intensity g. Both characteristic gravitational scales include the Planck constant, and thus they have a purely quantum nature.

Problem 2.24. *The bottom mirror is inclined relative to the horizontal plane by some angle α. Imagine that the mirror is sufficiently large in order to neglect any boundary side effects. Imagine also that the gravitational field is sufficiently uniform within the size of the mirror in order to neglect any gradient forces.*

Would a bouncing particle be settled in some quantum state above the surface of such a mirror?

If no, explain why.

If yes, describe the resulting quantum states.

In the limit of a hypothetical zero value of the Planck constant ($\hbar = 0$), both scales l_0 and t_0 (2.93, 2.94) would tend to zero as well, thus providing a typical classical behavior of such a hypothetical system with the zero value of the Planck constant. In the real world, this fundamental constant could not be changed, but the gravity strength and the gravitational mass of the object are values, which we could play with in order to optimize conditions needed to enhance quantum effects.

The same effect, as that obtained formally with the hypothetical zero Planck constant, would be achieved (see equations [2.93, 2.94]) if we would tend the value of a *gravity strength* in these formulas to *zero*. Thus, in order to enhance quantum effects in the motion of objects in a gravitational field, we would like to provide conditions with *weak gravitational fields*. In the opposite case of stronger gravitational fields, gravitational quantum effects would be suppressed.

The same effect of enhancing gravitational quantum effects would also be achieved if we tend the value of a gravitational mass to zero in the corresponding formulas. Thus, in order to enhance quantum effects, we prefer to use *light objects*. In order to profit simultaneously from the advantages of both these mentioned factors, we would like to observe the motion of light objects in weak gravitational fields.

Figure [2.3] illustrates in an artistic manner the difference in the behavior of objects with different mass, which are settled in low quantum states

Fig. 2.3 The lighter the higher.

in the gravitational potential above a reflecting surface. In any case the behavior is defined only by the value of mass and the value of the resulting efficient gravitational potential independent of its nature.

The motion of the whale with its huge mass is exhibiting a typical classical behavior because the characteristic scale of the corresponding quantum states would anyway be negligibly small.

The motion of the medium-mass extraterrestrial and the light butterfly is exhibiting quantum features, the spatial scale of a quantum motion of an object with a negligible mass in the gravitational potentials is relatively large.

It is curious to note how the two characteristic scales, l_0 and t_0, change as a function of the value of the inertial m and gravitational mass M. While the spatial gravitational scale l_0 depends on the value of inertial mass in the same way as it depends on the value of gravitational mass, this is not the case for the temporal gravitational scale t_0.

Thus, if we are interested in the verification of the equivalence principle in quantum mechanics, we should rather turn our attention to the characteristic spatial and temporal scale of quantum effects in the motion of objects in the gravitational field.

Could we measure practically these intrinsic quantum values of time and length of a falling particle?

We already know that the principal approach for "visualization" of any quantum scales is to use *interference* phenomena. A stationary state of a quantum particle, which is bouncing on a reflecting surface in a gravitational field, is a useful case of interference of the probability density waves, propagating in upward and downward directions. Such gravitational quantum states allow us to observe the spatial gravitational scale and the temporal gravitational scale characteristic for a bouncing particle.

Indeed, we have already discussed that the value of the spatial scale l_0 of a gravitational quantum effect could be extracted from analysis of the shape of density distribution to observe a quantum particle settled, for instance, in the ground gravitational state, while the value of the temporal scale t_0 could be extracted from analysis of the period of temporal oscillations of the probability density to observe a quantum particle settled, for instance, in a superposition of two lowest gravitational states.

We are going to discuss below the rich world of experiments with bouncing quantum particles and various practical means to measure the characteristic gravitational scales l_0 and t_0.

Now we are prepared to "test theoretically" the equivalence principle in quantum mechanics.

One could easily verify that the inertial mass m and the gravitational mass M can be "constructed" using the spatial scale l_0 and the temporal scale t_0:

$$m = \frac{\hbar t_0}{2 l_0^2}, \tag{2.96}$$

$$M = \sqrt{\frac{\hbar}{t_0 l_0 g}}. \tag{2.97}$$

Problem 2.25. *Verify the validity of the above expressions (2.96, 2.97).*

These two relations (2.96, 2.97) allow us to test the equivalence principle in quantum mechanics. As soon as we measure independently the values of characteristic gravitational length scale and characteristic gravitational time scale, l_0 and t_0, respectively, we could immediately calculate the values of inertial and gravitational masses, m and M, respectively, and compare them. If the equivalence principle is valid, they should be equal to each other.

Problem 2.26. *Verify that the gravitational spatial scale l_0 and the gravitational temporal scale t_0 are related to each other via the following expression:*

$$\frac{M}{m} = \frac{2l_0}{gt_0^2}. \tag{2.98}$$

One could easily recognize that in the case of equality of the two masses, $m = M$, the above expression (2.98) is converted into the well-known expression for the classical time-of-fall: $t_0 = \sqrt{2l_0/g}$. We see that the relation between the quantum gravitational temporal scale and the quantum gravitational spatial scale is *exactly the same*, as it is in classical mechanics, as long as we assume *validity of the equivalence principle* in quantum mechanics.

Problem 2.27. *The inertial mass m of an object could be measured in an accelerated frame of reference, which is characterized by a flat artificial gravity acceleration (the force lines are parallel to each other).*

The gravitational mass M of this object could be measured in a gravitational field, which is always characterized not only by a mean gravitational acceleration but also by some gradient forces (the force lines are not parallel to each other, they are directed to a common point).

A quantum object is always nonlocal, which means that it is sensitive not only to the strength of gravitational or artificial force in a given point but also to the strength of the force in a vicinity of this point.

Taking into account all these statements, could you explain the meaning of the equivalence principle in quantum mechanics?

We have just introduced the equivalence principle in quantum mechanics and we enjoy its beauty. However, we are already shaking the ground under this structure. Well: "We are trying to prove ourselves wrong as quickly as possible, because only in that way can we find progress" (Richard Feynman).

Thus, neglecting the gravitational gradient forces, a direct consequence of validity of the equivalence principle in quantum mechanics is the universality of the ratio between space and time; this ratio does not include the mass of a quantum particle. At the same time quantum motion in a gravitational field, understood in terms of the evolution of the probability density to observe the particle at a certain height at a certain moment of time, depends directly on the mass of the quantum particle!

2.4 Συνοψις

Our treatise on quantum motion of small and light particles, like atoms, neutrons and electrons, started with two counterintuitive statements:

i) In the micro-world, a particle motion could no longer be described using a *position* $\vec{r}$ of the particle at a given moment t; instead it should be described using a complex-valued field, which is also called a *wave-function* $\Psi(\vec{r}, t)$.

ii) Physical reality is associated with the *modulus square* of this field; namely, the modulus square of a wave-function $|\Psi(\vec{r}, t)|^2$ is equal to the *probability density* to observe the particle in the vicinity of a position $\vec{r}$ at a time t.

As a physical observable value is the modulus square of a wave-function, but not a wave-function itself, this statement means in particular that all wave-functions, which differ from each other with an unitary factor $exp(iS(\vec{r}, t))$ with any real-valued *phase* $S(\vec{r}, t)$, are physically equivalent.

The quantum law of motion, known as the *Schrödinger equation*, is a differential equation that allows us to calculate a wave-function in any location in space at any instant of time:

$$i\hbar\frac{\partial\Psi(t,\vec{r})}{\partial t} = \left[-\frac{\hbar^2}{2m}\nabla^2 + U(\vec{r})\right]\Psi(t,\vec{r}). \qquad (2.99)$$

The fundamental constant $\hbar$ with the dimension of action, known as the *Planck constant*, establishes the relation between a typical spatial scale of a wave-function and a momentum of the corresponding particle; it also establishes the relation between a typical temporal scale of a wave-function and the energy of the corresponding particle:

$$p\lambda = 2\pi\hbar, \qquad (2.100)$$

$$ET = 2\pi\hbar. \qquad (2.101)$$

The *principle of relativity* applied to the Schrödinger equation imposes that the equation should keep its form (should stay *covariant*) being written in any inertial frame of reference. This condition could be met only if we substitute a wave-function with its equivalent:

$$\Psi_1 = \exp(iS(\vec{r},t))\Psi_2 \qquad (2.102)$$

simultaneously with the coordinate transformations. Here $\hbar S(\vec{r},t)$ is a *classical action* of a freely moving particle.

The essence of quantum motion, thought of as a time evolution of a probability density distribution, is an *interference* phenomenon. Interference means that a solution of the Schrödinger equation $\Psi(\vec{r},t)$ could be expressed as a sum (usually as an infinite sum) of wave-functions of a certain type (for example, as a sum of plane waves, which are the solutions of the free-motion Schrödinger equation).

This sum results in *enhancing* the wave-function in some spatial points and in *reducing* it in other points, so that the observed probability density $|\Psi(\vec{r},t)|^2$ has maxima and minima in different spatial locations. In general, the locations of maxima and minima of the probability density evolve. A special case of such an evolution is a motion of a wave-packet; this type of quantum motion is the closest analog of a motion of a classical object.

It is instructive to study the *free motion* of an initially shaped wave-packet $\Psi_0(z)$. Its evolution could be predicted analytically:

$$\Psi(z,t) = \int_{-\infty}^{+\infty} G(z,z',t)\Psi_0(z',t=0)dz', \qquad (2.103)$$

$$G(z,z',t) = \sqrt{\frac{m}{2\pi i\hbar t}} \exp\left(iS(z,z',t)/\hbar\right), \qquad (2.104)$$

$$S(z,z',t) = \frac{m}{2t}(z-z')^2. \qquad (2.105)$$

A solution of the Schrödinger equation (2.99) with zero potential $(U(z) = 0)$ is expressed as a sum (as an integral) of solutions for the evolution of an individual point-like source with an amplitude $\Psi_0(z)$ located at a height z. The result of summation for the particular case of a Gaussian wave-packet could be calculated in a closed form:

$$|\Psi(x,\tau)|^2 = \frac{1}{\sqrt{\pi}\sigma\sqrt{1+\hbar^2t^2/(m^2\sigma^4)}} \exp\left(\frac{(z-z')^2}{2\sigma^2(1+\hbar^2t^2/(m^2\sigma^4))}\right).$$
$$(2.106)$$

One could easily derive what happens to a wave-packet with the initial velocity v by using the principle of relativity and the transformation rule (2.102) for the wave-function in a new inertial frame of reference:

$$|\Psi(x,\tau)|^2 = \frac{1}{\sqrt{\pi}\sigma\sqrt{1+\hbar^2t^2/(m^2\sigma^4)}} \exp\left(\frac{(z+vt-z')^2}{2\sigma^2(1+\hbar^2t^2/(m^2\sigma^4))}\right).$$
$$(2.107)$$

This result could be interpreted as the interference of different terms in the sum (2.103). This interference cancels out the resulting sum everywhere except for the vicinity of the classical trajectory $z' = z + vt$. Such a localization of the center of a wave-packet on the classical trajectory is a very important consequence of a close relation between quantum and classical mechanics.

Formally, this relation follows from the fact that the phase of an oscillating exponent is a classical action. According to the stationary phase method of calculating the corresponding integral, the contribution of those terms is dominant, which provides the extremum of a phase $S(z, z', t)$.

However, these terms correspond to coordinates of the *classical trajectory* according to the *least action principle*!

Another intriguing example is *free fall* of a quantum particle.

A deep relation between the principle of relativity and the equivalence principle could be illustrated by means of analyzing a quantum fall in a freely falling frame of reference (which is a classical object!). The transformation of coordinates $z' = z - gt^2/2$ is combined with the transformation of the wave-function according to equation (2.102), where $S(z, t)$ is the classical action for a freely falling object.

We get, as a *consequence* of the equivalence principle $(m = M)$, that an equation for Ψ_2 (2.102) is the equation for a freely moving particle. This consequence allows us to get the probability density of a quantum particle in an inertial frame of reference:

$$|\Psi_1(x, \tau)|^2 = \frac{1}{\sqrt{\pi}\sigma\sqrt{1 + \hbar^2 t^2/(m^2\sigma^4)}} \exp\left(\frac{(z - z' + gt^2/2)^2}{2\sigma^2(1 + \hbar^2 t^2/(m^2\sigma^4))}\right).$$

$$(2.108)$$

Thus, a free fall wave-function could be transformed into a free motion wave-function universally (i.e. simultaneously for particles of different mass). This fact shows that the concept of free fall is a *generalization* of the concept of free motion. This is the essence of the equivalence principle, which substitutes the idea of a *dynamical interaction* (Mgz) in the case of a gravitational field with the idea of *space-time properties* (the coordinate and wave-function transformations), which is *universal* for any particle.

Let us mention that in both cases (free motion and free fall) the center of a wave-packet moves along the classical trajectory. However, a new quantum effect seen here is a *spreading* of a wave-packet with time. This spreading is faster for smaller initial localization σ of the wave-packet.

Such a spreading could be explained in terms of interference of waves with different momenta, i.e in terms of a momentum distribution associated with a given wave-function:

$$\Psi(z,t) = \int_{-\infty}^{\infty} C(p) \exp\left(ipz/\hbar - ip^2/(2m\hbar)t\right) dp, \qquad (2.109)$$

$$C(p) = \frac{\hbar}{(2\pi)^{1/2}} \int \Psi(z,0) \exp(-ipz/\hbar) dz. \qquad (2.110)$$

The spread of the momentum distribution Δp is related to the spatial localization σ via the uncertainty relation $\Delta p \sigma \leq \hbar/2$; the minimum is achieved for a Gaussian-type wave-packet.

The spreading of an initially localized wave packet in time is due to interference of plane-waves with different momenta. They contribute to the wave-packet $\Psi(z, t = 0)$, each plane wave with an amplitude $C(p)$, in such a way that the integral (2.109) cancels out everywhere except for the localized area with the width σ. The evolution of the wave-packet is encoded in a time-dependent phase $\exp\left(-ip^2/(2m\hbar)t\right)$, which is different for waves with different momenta. This *dephasing* results in the wave-packet spreading.

Thus, quantum motion itself is an interference phenomenon. This type of motion covers a much broader manifold of cases than classical motion along a trajectory. Only in certain cases (like a motion of a wave-packet center) it looks similar to classical motion. In ultimate cases, quantum motion has nothing to do with its classical analog.

An example of such an ultimate case is a stationary state, or a state with a certain value of energy:

$$E\varphi_E(z) = \left[-\frac{\hbar^2}{2m}\frac{d^2}{dz^2} + U(z)\right]\varphi_E(z). \qquad (2.111)$$

This time-independent Schrödinger equation should satisfy a boundary condition, which usually consists of limiting the space available for the particle motion like $\varphi(z \leq 0) = 0$. The latter condition is equivalent to the following equality $\varphi(z = 0) = 0$.

The spatial distribution in a wave-packet composed of only *one* stationary state, does not evolve:

$$|\Psi(z,t)|^2 = |\varphi(z)|^2. \qquad (2.112)$$

Quantum motion of a wave-packet, which is a combination of *several* stationary states, exhibits a wide variety of temporal dependencies, from harmonic oscillations in the case of two states to a classical-type motion along a trajectory in the case of many states with close energies contributing to a wave-packet.

$$|\Psi(z,t)|^2 = \sum_i |\varphi_i(z)|^2 + 2\sum_{i>j} \varphi_i(z)\varphi_j(z) \cos\left[(E_i - E_j)t/\hbar\right]. \quad (2.113)$$

Our interest is focused on gravitational states of a quantum particle above a reflecting surface:

$$E\varphi_E(z) = \left[-\frac{\hbar^2}{2m}\frac{d^2}{dz^2} + Mgz\right]\varphi_E(z), \quad (2.114)$$

$$\varphi(0) = 0. \quad (2.115)$$

There is *a clock and a ruler* intrinsically embedded in these states. Indeed, spatial and temporal units could be defined as a combination of the mass, the gravity field intensity and the Planck constant:

$$l_0 = \sqrt[3]{\frac{\hbar^2}{2mMg}}, \quad (2.116)$$

$$t_0 = \hbar/\varepsilon_0 = \sqrt[3]{\frac{2m\hbar}{M^2g^2}}. \quad (2.117)$$

The form of the wave-function and the energy of a gravitational state, expressed in dimensionless variables $x = z/L_0$, $\lambda_n = E_n/\varepsilon_0$, is the following:

$$\varphi_n(x) = \frac{\text{Ai}(x - \lambda_n)}{|\text{Ai}'(0)|}, \quad (2.118)$$

$$\text{Ai}(-\lambda_n) = 0. \quad (2.119)$$

The spatial density distribution of a gravitational state has a well-defined *gravitational length scale* l_0, which depends on both inertial and gravitational masses (2.116). The energy level spacing has a well-defined *gravitational energy, and thus temporal scale* t_0, which also depends on inertial and gravitational masses (2.117). These relations allow us to extract the values of inertial and gravitational masses from an experiment with a bouncing quantum particle:

$$m = \frac{\hbar t_0}{2l_0^2}, \tag{2.120}$$

$$M = \frac{\hbar}{t_0 l_0 g}. \tag{2.121}$$

$$\tag{2.122}$$

In quantum physics, gravitational spatial and temporal scales are related to each other like they are related in classical mechanics, namely:

$$\frac{M}{m} = \frac{2l_0}{gt_0^2}. \tag{2.123}$$

The equivalence principle imposes that $t_0^2 = 2l_0/g$.

Thus, gravitational states provide us an access to such an intriguing quantity as the gravitational mass as well as to the *quantum equivalence principle* test.

In the following chapter we are going to discuss challenging experiments with matter and antimatter quantum particles bouncing on material surfaces in a gravitational field.

Chapter 3

Bouncing Neutrons

3.1 Ultracold Neutrons and Challenge of Gravity

An idea to observe experimentally a *quantum particle*, bouncing on a horizontal material surface in the gravitational field of the Earth, seems to be very attractive.

First, such an observation would provide us the *first case* of quantum states of matter in a gravitational field measured experimentally. We are aware of examples of quantum states of matter in other fundamental fields: for instance, atoms are quantum states of electrons in the electromagnetic field of the nucleus, and nuclei are quantum states of nucleons in their strong nuclear field. However, gravity had been missing in this list until recent times.

Second, such an observation is an enormous *experimental challenge* in itself, and thus prompting us to solve this problem.

A good illustration for the last statement is a phrase written by Brian Hatfield in the famous "Feynman Lectures on Gravitation" published as recently as 1995:

> Let us consider another possibility, an atom held together by gravity alone. For example, we might have two neutrons in a bound state. When we calculate the Bohr radius of such an atom, we find that it would be 10^8 light year, and that the atomic binding energy would be 10^{-27} Rydbergs.
>
> There is then little hope of ever observing gravitational effects on systems which are simple enough to be calculated in quantum mechanics.[1]

---

[1] Feynman, R. P., Morinigo, F. B., and Wagner, W. G. (1995). *Feynman Lectures on Gravitation*, Addison-Wesley, USA, p. 11.

Third, a realization of this idea would provide us with a quantum system with such behavior in a gravitational field that we could *predict precisely*. Moreover, as far as the mirror could be considered as an ideal impenetrable barrier, the behavior of such a quantum system depends *only on the gravitational field*. We are going to show in the following two chapters that the latter condition is not an exotic case but a very natural result, for instance for neutrons, and even for atoms and antiatoms.

Fourth, such a simple system with parameters, which depend only on a gravitational field, is an *excellent instrument* for various studies both in fundamental science and in practical applications, which we are also going to mention briefly in the rest of the book. The principle advantages are extreme fragility of gravitational quantum states associated with their extremely low energy, as well as simultaneous weakness of systematical false effects in a properly designed experiment.

Fifth, such a "textbook" system is very convenient for educational purposes.

And the reader could probably continue and continue the list of various additional motivations.

However, for many years any practical realization of such an experiment was considered as a hardly ever possible adventure.

One reason for these doubts is well justified by the fact that gravity is much weaker than any other fundamental interaction. The *weakness of gravity* means that their effects would be completely washed out by much stronger effects of other interactions. The most dangerous competitor in this sense is the electromagnetic interaction because it is a long-ranged force, just like gravity is; however, it is many orders of magnitude stronger than gravity.

In order to get an idea about typical strengths of gravitational and electromagnetic interactions in simplest systems, perform the following exercise.

Exercise 3.1. *An electron in a hydrogen atom is attracted to the nucleus (i.e. just to a proton) due to the existence of their electric charges. However, the electron and the proton are also attracted to each other due to the gravitational interaction.*

Estimate how many times the first interaction is stronger than the other interaction.

This comparison imposes that the best candidate for such a quantum probe of gravity is an *electrically neutral particle*. However, even a neutral elementary particle is composed of electrically charged components – quarks – and thus it participates in electromagnetic interactions, for instance due to the existence of its magnetic (dipole and other) moment(s). An electrically neutral atom is a less compact system than a neutron, therefore it is more affected by any residual electromagnetic fields to be strongly suppressed.

Another challenge, which makes an observation of bouncing and even free fall of a quantum particle very complicated, results from *large typical velocities* of quantum particles, which we "have at hand". In particular, the mean velocity of a lightest atom, the hydrogen atom, or the mean velocity of a neutron is as large as about 2200 m/s at the ambient temperature. Such a particle could raise in the gravitational field of the Earth (in vacuum) to the height of approximately 250 km!

This estimation of a typical velocity and a typical spacial scale shows that any observation of gravitational states of quantum particles at the ambient temperature is absolutely impractical. The particles should be cooled down to *ultra-low temperatures*, at which the quantum properties of their motion in the gravitational field of the Earth would become dominant on one hand, and an experiment with such quantum particles would become feasible in typical laboratory conditions on the other hand.

Exercise 3.2. *Calculate the mean effective vertical velocity of a neutron settled in the lowest gravitational quantum state.*

As the velocity and the energy of a quantum particle has to be that low, and the difference of energies of neighboring quantum states is that low as well, would it possible to resolve experimentally the neighboring quantum states? One important condition follows from the uncertainty principle, which states that the *time of observation* of a particle in a quantum state has to be at least as long as the characteristic gravitational time scale, or better it should be much longer than that.

Exercise 3.3. *Calculate the characteristic gravitational time scale for the quantum motion of a neutron in the gravitational field of the Earth.*

Evidently, most kinds of neutral elementary particles have shorter life-times and therefore they could not be used. However, a long enough lifetime is a necessary condition but not a sufficient condition. Also, the energy of a particle should not change at the gravitational time scale. While any *inelastic interaction* with the gravitational field is quite ruled out[2], the problem of whether the interaction with a mirror is sufficiently elastic is less evident and we would have to consider it below in more detail.

The degree of (in)elasticity of interaction with a mirror would depend on the kind of particle and mirror chosen, as well as on the nature and param-eters of the interaction of the particle with the mirror. One could expect that it is natural to search for a laboratory probe for gravity studies among atoms. The reason is twofold: i) we have a lot of atoms at hand; ii) there are developed methods to cool atoms down to the extreme temperatures needed.

However, the first particle (and up to this date the only one), with which gravitational quantum states were *experimentally observed*[3], appeared to be the neutron.

Exercise 3.4. *Estimate the characteristic scale of effective temperatures of quantum particles with the mass equal to one atomic unit (such as neu-trons, hydrogen or antihydrogen atoms) bound in low quantum states in the gravitational field of the Earth.*

In order to understand why neutrons appeared to be so special for this particular purpose, let us describe briefly their principle properties that are important for the following presentation and the history of investigations of/with neutrons. The neutron is a neutral particle with a mass approxi-mately equal to that of a proton. The neutron was discovered[4] by Sir James

---

[2]Pignol, G., Protasov, K. V., and Nesvizhevsky, V. V. (2007). A note on spontaneous emission of gravitons by a quantum bouncer, *Class. Quant. Grav.* **24**, p. 2439.

[3]Nesvizhevsky, V. V., Boerner, H. G., Petukhov, A. K., Abele, H., Baessler, S., Ruess, F. J., Stoeferle, Th., Westphal, A., Gagarski, A. M., Petrov, G. A., and Strelkov, A. V. (2002). Quantum states of neutrons in the Earth's gravitational field, *Nature* **415**, p. 297.

[4]Chadwick, J. (1932). Possible existence of a neutron, *Nature* **129**, p. 312.

Chadwick in 1932 in the early years of exploration of atomic nuclei. It was a nontrivial discovery.

The neutron turned out to be the first discovered neutral massive particle. Before then only electromagnetic interactions and gravity were known. The role of neutrons as building blocks of matter was unclear. The idea that the neutron is an important constituent of atomic nuclei and is not composed of known elementary particles was first put forward by Ambarzumian and Iwanenko[5]; this idea required an assumption about a new type of interaction in nature.

This interaction is a so-called "strong force", which keeps the atomic nuclei stable. A characteristic property of this force is its short-range character; a typical radius of the strong force interaction, i.e. the distance beyond which the interaction strength significantly decreases, is of the order of 10^{-13} cm (or 1 Fermi). This distance is much smaller than a typical electromagnetic size of an atom, or an interatomic distance, the latter is equal a few times to the atomic Bohr radius, i.e. $0.5 \cdot 10^{-8}$ cm (or 0.5 Angstrom).

Thus, when a (electrically charged) proton is interacting with a solid body, atoms in the body and nuclei in the atoms could be compared with a thicket. When a (electrically neutral) neutron of a typical energy and wavelength is interacting with a solid body, nuclei could be compared with rare trees in an African savannah (we talk about nuclei in atoms in the solid body but not about atoms because the interaction of neutrons with electrons in atoms is relatively weak).

This difference in electromagnetic interactions (while strong interactions of neutrons and protons are similar) illustrates the ability of neutrons, in contrast to protons and electrons, to *penetrate deep* inside solid bodies. Due to this amazing property, even rather slow neutrons could penetrate deep inside bodies and inside atomic nuclei, and therefore they could cause nuclear reactions. In contrast, protons are reflected from nuclei, and could penetrate inside only if their energy is high enough.

That is why slow neutrons play, for instance, a key role in practical atomic-power applications, and at the same time that is how neutrons can be detected. Typical fragments of nuclear reactions caused by neutrons are high-energy charged particles, which are capable of ionizing the detector

---

[5]Ambarzumian, V., and Iwanenko, D. (1930). Les electron inobservable et les rayons, *Compt. Rend.* **190**, p. 582.

matter and thus inducing a photo- or electrical response in a detector. Thus, fragments of nuclear reactions signal about the presence of neutrons, which cause these reactions.

Neutrons are an important constituent of matter. Thus, more than a half of the mass of your body, as well as more than a half of the mass of surrounding objects, is composed of neutrons. This is so because the mass of electrons in atoms is negligible, and the number of neutrons in nuclei is usually larger than the number of protons, while the mass of a neutron is about equal to the mass of a proton. Although neutrons are everywhere, nuclear reactions are the only source of *free* neutrons that we have at hand.

The reason for this particular role of nuclear reactions is that a free neutron is an *unstable* particle with the relatively short lifetime of approximately 880 s. As soon as a neutron is released it starts decaying; the decay rate is defined by another fundamental interaction: weak interaction. On the other hand, neutrons inside nuclei are usually stable and could stay there safe until the event of some nuclear reaction, caused by a collision with an external particle.

A typical energy of most neutrons, which are released in nuclear reactions, is of the order of a few MeV; a corresponding velocity is of the order of dozens of thousands of kilometers per second. After multiple elastic (in the center-of-mass frame of reference) collisions with atomic nuclei in a moderator medium, neutrons transfer their kinetic energy to these atomic nuclei and thus are "thermalized", i.e. they reduce their energy to the values typical for the energy of atomic thermal motion[6].

A corresponding mean neutron velocity at the ambient temperature is about 2200 meters per second. Such neutrons are called *thermal neutrons*. There is no chance to use thermal neutrons as particles, which could exhibit quantum bouncing in the gravitational field of the Earth above a material surface. Such neutrons are too fast. We have already estimated above that thermal neutrons would raise in the gravitational field to incredible heights incompatible to the gravitational quantum scale.

Problem 3.1.   *Estimate the number of collisions needed for a neutron with the energy of 2 MeV to cool down in liquid hydrogen medium to the typical energy of thermal neutrons.*

---

[6]Anderson, H., Fermi, E., and Szilárd, L. (1939). Neutron production and absorption in uranium, *Phys. Rev.* **56**, p. 284.

Note: Take into account that in the considered range of energies neutrons are scattered elastically on hydrogen atoms in the center-of-mass frame of reference and thus the process of collision is totally characterized by two conservation laws, the conservation of energy and the conservation of momentum.

Note: Take also into account that the angular distribution of scattered neutrons is isotropic in the center-of-mass frame of reference and then ... you will derive immediately the theory of neutron thermalization in reflectors of nuclear reactors.

For gravitational quantum experiments, neutrons should be somehow cooled down to even smaller temperatures. Below the energy range of thermal neutrons, one distinguishes several other energy groups, respectively: cold neutrons (CNs), then very cold neutrons (VCNs), and finally *ultracold neutrons (UCNs)*. The name of UCNs is justified by the fact that their typical energy is as low as about a hundred neV, their typical velocity is a few meters per second and their typical temperature is as low as 1 mK.

The thermalization due to billiard-ball-like collisions of neutrons with nuclei described above is not suitable in this range of parameters because the neutron wavelength starts to be comparable to a characteristic interatomic distance and a *coherent simultaneous interaction* of a neutron with several nuclei could no longer be neglected. Some exotic tricks could be probably applied in the future to extend the range of relevant neutron energies to lower values[7], but this is definitely not a question for the near future.

Exercise 3.5. *Coherent simultaneous interaction of a neutron with several nuclei in a medium starts to be significant as far as the neutron wavelength starts to be comparable to a characteristic interatomic distance in the medium.*

Estimate the corresponding characteristic parameters of the neutron for a material of your choice:

- the wavelength

- the energy

---

[7]Nesvizhevsky, V. V. (2002). Interaction of neutrons with nanoparticles, *Phys. At. Nucl.* **65**, p. 400.

- the velocity

- the temperature

- the raising height in the Earth's gravitational field.

Up to now people rather select a tiny low-energy fraction in a much broader equilibrium neutron spectrum, or use so-called super-thermal neutron converters, which provide nonequilibrium down-scattering of faster neutrons in a single scattering event. In spite of major worldwide efforts to increase densities and fluxes of UCNs for physical experiments, available maximum densities are ridiculously small: a few tens of UCNs per cubic centimeter, barely higher than the density of interstellar gas.

However, one *remarkable feature of UCNs* justifies all these efforts: neutrons cooled down to UCN temperatures are totally reflected from surfaces of most materials and thus they could be trapped in closed vessels. Thus they could also bounce on material surfaces for a long time. A particularly important parameter of such neutron reflection from a surface is that it is elastic in a majority of cases. Only a small fraction of UCNs per reflection is absorbed or scattered inelastically.

UCNs with such a low typical temperature as one or a few milliKelvin could raise in the gravitational field of the Earth up to the height of only a few meters. Nevertheless such values are still much higher than the characteristic gravitational quantum scale, which is measured in micrometers. Nearly six orders of magnitude in terms of energy, or respectively in terms of temperature, are still missing. Nevertheless, we are approaching the right scale!

From the point of view of classical theory, the total reflection of UCNs from a material surface is absolutely counterintuitive. Indeed, the main property of neutrons, which we have mentioned above, is their high penetrability through materials, which could be understood as a result of their electrical neutrality and, as a consequence, a small neutron–nuclei interaction cross section. Naively speaking, in classical physics neutron is a small particle, which could easily penetrate between the nuclei of media.

Could be the total reflection of UCNs from surfaces explained, say, by an enormous increase of the neutron–nuclei interaction cross-section at low UCN energies? No, this assumption is in sharp contradiction with known facts about strong forces. The phenomenon of total reflection of UCNs from most materials should not be interpreted in the framework of classical physics; in contrast, it should only be explained in terms of wave properties of neutrons.

The quantum mechanical explanation is based on the interference of waves, which are simultaneously scattered on an ensemble of atomic nuclei in the medium. The propagation of UCN wave in the medium could be described in terms of an effective potential[8], a so-called *neutron optical potential*, which is generated by the nuclei in the medium. For the interaction of UCNs with a homogeneous medium this optical potential represents a step function. Let us analyze this fact qualitatively.

To start with, let us return to the analogy already used that when a neutron is interacting with some matter then nuclei in the matter could be compared with rare trees in an African savannah. This is true if the neutron velocity is relatively large, or in other words if the neutron wavelength is smaller than the distance between nuclei (atoms) in the matter. However, a slow enough neutron is "distended" to the size larger than the interatomic distance and does not fit in-between atoms any longer.

First, let us outline an important general fact from the theory of scattering of waves. It consists of emerging of a *spherically symmetrical scattered wave* (it is called an S-wave) when an incoming plane wave interacts with an obstacle with a size much smaller than the wave length. A typical wave length of a UCN is 50–100 *nm*, much larger than a typical range of the neutron–nuclei interaction potential, as well as much larger than a typical interatomic distance.

These observations mean that each nucleus becomes a source of a secondary spherically symmetric wave while a UCN wave propagates through a layer of atoms in a medium. The amplitude of this scattered wave is defined by the amplitude of an incoming neutron wave on one hand and by the neutron–nuclei interaction intensity on the other. Thus, a neutron wave-function $\Psi(\vec{R})$ in a given point $\vec{R}$ is a superposition of incoming and scattered waves:

$$\Psi(\vec{R}) = \Psi_0(\vec{R}) - f \sum_i \frac{exp(ikr_i/\hbar)}{r_i} \Psi(\vec{R}_i). \tag{3.1}$$

Here k is the modulus of the neutron momentum, $\vec{R}_i$ is the position of i-th nucleus, $r_i = |\vec{R} - \vec{R}_i|$ is the distance from a given nucleus to the

---

[8]Fermi, E., and Marshall, L. (1947). Interference phenomena of slow neutrons, *Phys. Rev.* **71**, p. 666.

observation point and f is a factor, which is responsible for the intensity of neutron–nuclei scattering, the so-called scattering amplitude.

Second, we are going to profit from the fact that a neutron wave is *weakly perturbed* by each scattering nucleus. The weak perturbation means that we could choose such a thin layer of atoms with the depth Δz that neutron scattering in this layer would only slightly modify the incoming wave. This assumption means that one could replace a precise wave-function $\Psi(\vec{R}_i)$ with the undisturbed wave-function $\Psi_0(\vec{R})$ in the last term in equation (3.1), describing the sum of scattered waves.

We also take into account that the incoming wave could be expressed as a plane wave with a given momentum $\vec{k}$ which is orthogonal to the layer plane, and thus $\Psi_0(\vec{R}) = \exp(ikz)$. We also substitute summation over positions of nuclei by integration of an over-effective continuous medium with the density n (which we consider to be constant within our thin layer). This substitution is possible because the integrand changes significantly over the wavelength $\hbar/k$, which is much larger than the distance between neighboring nuclei.

Thus the expression for the neutron wave-function could be formulated as follows:

$$\Psi(\vec{R}) = \exp(ikz) - 2\pi f \Delta z n \exp(ikz_0) \int_0^\infty \frac{\exp(ikr(\rho)/\hbar)}{r(\rho)} \rho d\rho. \qquad (3.2)$$

The integral in the above expression (3.2) could be calculated explicitly. As far as we take into account the equality $r(\rho) = \sqrt{(z-z0)^2 + \rho^2}$, it turns into the following expression:

$$\int_0^\infty \frac{exp(ikr(\rho)/\hbar)}{r(\rho)} \rho d\rho = \int_{z-z_0}^{\rho_\infty} \exp(ikr)dr = \frac{\exp(ik\rho_\infty) - \exp(ik(z-z0))}{ik}.$$
$$(3.3)$$

The integration is performed here up to some very large value of a radial coordinate $\rho_\infty \gg \hbar/k$. If one averages this result over a very narrow range of momenta $dk \sim \hbar/\rho_\infty$ around the mean value k, then fast oscillations of the factor $\exp(ik\rho_\infty)$ would vanish the corresponding term. Thus the value of the integral is equal $i\exp(ik(z-z0))/k$. Finally, for the neutron wave, which passed through a thin layer of atoms with a depth $\Delta z - z - z_0$ and a density n, we get:

$$\Psi(\vec{R}) = \exp(ikz/\hbar)(1 - 2i\pi f\hbar(z - z_0)n/k). \qquad (3.4)$$

This wave-function (3.4) depends only on z-coordinate, and this result follows immediately from the symmetry of the problem.

In other words a neutron wave, which falls normally to the surface of a medium, is *not deflected*. This conclusion contradicts a classical picture of stochastic wandering of a small ball, which is scattered on a random ensemble of scattering centers. Instead, a neutron wave propagates in an effective uniform field. To find this effective potential let us substitute the wave-function (3.4) into the Schrödinger equation with a uniform potential U, which acts only within a thin layer.

Exercise 3.6. *Verify that the wave-function (3.4) satisfies the stationary Schrödinger equation with the following potential:*

$$U = \frac{2\pi\hbar^2 fn}{m}. \qquad (3.5)$$

Note: Neglect second-order terms, which characterize neutron scattering in the thin layer.

The remarkable result, which is contained in the solution of the above exercise, establishes that a neutron wave "feels" not individual nuclei, but the whole ensemble of scattering centers with a density n and an individual scattering amplitude f. This conclusion means that a neutron wave undergoes an action of an effective optical potential (3.5). The fact that a neutron wave is scattered "simultaneously" on many nuclei is clearly seen in expression (3.3).

The integral, which is responsible for the contribution of secondary scattered waves, converges while it is calculated over a distance comparable with a few times the neutron wavelength $\hbar/k$, or larger than that. That is why the validity of the optical potential approximation is guaranteed due to the fact that a neutron wavelength is much larger than a characteristic interatomic distance. This condition is satisfied for UCNs with a high accuracy.

Let us turn to the question of how individual properties of scattering of a neutron on nuclei, which compose the medium, contribute to the resulting optical potential of this medium. For all known isotopes of all nuclei, a typical absolute value of the neutron–nucleus scattering amplitude f, which enters into the expression for an optical potential (3.5), is of the order of a few Fermi (to remind this length unit: 1 Fermi equals 1 femtometre, or 10^{-13} cm).

Generally speaking, an optical potential is a *complex value*. Consider its imaginary and real parts.

An *imaginary part* of the scattering amplitude is negative. It corresponds to the neutron absorbtion in a nucleus. For a UCN with its extraordinary low energy and with a temperature that is much larger than the medium temperature, any inelastic scattering in the medium usually results in losing the UCN from its initial energy range, and thus to losing the UCN from its material trap. Therefore, all kinds of inelastic scattering are added to the neutron absorbtion.

A *real part* of the scattering amplitude could be either positive (for most isotopes of most nuclei) or negative. The two signs of the real part mean that a corresponding optical potential could be of both signs as well, i.e. it could be *repulsive* (the positive sign) or *attractive* (the negative sign). Usually the absolute value of the real part of the scattering amplitude is much larger than the absolute value of the imaginary part of the scattering amplitude.

(*Note*: If a medium is composed of nuclei of different kinds, *and* if the mixture of these nuclei is uniform on a spatial scale a few times smaller than the neutron wavelength [and at larger scales as well] then one could discuss a *mean* optical potential with a mean real and imaginary parts of the scattering amplitude. The above-mentioned uniformity of nuclei distribution is important as otherwise the matter would be seen by the neutron as separate domains with different potentials.)

Exercise 3.7. *Calculate the (complex) values of optical potential for the following materials:*
 i) diamond

ii) copper

iii) titanium.

Solving the above exercise allows us to illustrate and better understand the remarkable fact, which was stated in the origin of our present discussion, namely the phenomenon of total reflection of a UCN from most materials. The essence of this phenomenon is as simple as follows: Most materials are "perceived" by an ultracold-neutron-wave as a repulsive optical potential with a value of a few hundred nanoelectron-volts (neV).

As the typical kinetic energy of a UCN is smaller than that, the UCN is totally reflected from such an optical potential (which emerges due to coherent action of many nuclei within a volume in the medium with dimensions, which are of the order of the neutron wave length). There is no accidental coincidence of these two typical values: UCN energy and optical potential. In fact, UCNs are defined as those neutrons, which have such small energy that they could be reflected from typical optical potentials!

The value of optical potential of a medium is sometimes called the *critical energy* of the medium. It is expressed in energy units. However, one often also uses special units of energy, which are convenient for analyzing experiments with UCNs: both the UCN energy and the critical energy of a medium could be expressed in length units (centimeters). This definition is based on the observation that a UCN with the energy of 1 neV could raise in the gravitational field of the Earth to a height of approximately 1 cm.

Exercise 3.8. *Verify the above statement that a UCN with the energy of 1 neV could raise in the gravitational field of the Earth to the height of approximately 1 cm, and calculate precisely the relation between energy and length units for UCNs.*

Exercise 3.9. *Calculate the value of critical energy of sapphire, and express it in neV and in cm.*

Repulsive properties of a medium for a UCN could be characterized also in terms of a so-called *critical velocity* of the medium, which is defined as the velocity of a neutron with the kinetic energy equal to the value of the repulsive optical potential of the medium. UCNs with velocities larger than that could penetrate deep inside the medium, provided that the angle of incidence is large enough. In fact only the component of UCN velocity perpendicular to the surface is relevant.

Problem 3.2. *Calculate the typical depth of penetration of a UCN with the velocity of 1 m/s inside:*

   i) diamond

   ii) copper

   iii) titanium.

   Note: Consider that neutrons fall normally to the interface between the vacuum and the medium.

   Note: Take into account that the wave-function is a continuous function at the interface.

The same coherent action of many scattering centers in a large volume in a medium (compared to the volume occupied by one atom) explains another nontrivial property of UCNs: their reflection from a solid or liquid surface with a temperature of many orders of magnitude larger than the characteristic temperature of UCNs turns out to be *elastic* with an overwhelming probability, i.e. the energy is not transmitted from the "hot" atoms of a medium to a UCN.

The physical reason for this extraordinary feature[9] is that thermal oscillations of individual atoms in a solid or liquid medium, averaged over a large region with the dimensions of the order of a neutron wavelength, result in a vanishing average contribution.

The beauty and simplicity of the quantum mechanical explanation of the counterintuitive fact of neutron reflection from media was established by Enrico Fermi. Even the optical neutron medium potential is sometimes called the Fermi potential (or Fermi effective potential). Surprisingly, Enrico Fermi doubted the practical feasibility of neutron storage in material vessels, while it is an immediate consequence of the results on optical potential, discussed above.

---

[9]Barabanov, A. L., and Belyaev, S. T. (2000). Multiple scattering theory for slow neutrons (from thermal to ultracold), *Europ. Phys. J. B* **15**, p. 59.

Such a possibility was pointed out in a short paper written by Yakov Zeldovich[10]. A year later, Vasily Vladimirsky proposed to store UCNs in magnetic traps (due to the neutron magnetic moment) and also to use converters for the production of UCNs and neutron guides to transport UCNs from such an internal converter to external experiments[11]. In the same publication he used the term "ultracold neutrons" for those neutrons that could be trapped in material vessels.

The first experimental observation of trapped UCNs became possible due to the reflection of UCNs from the walls of a copper neutron guide with the length of 10 m and the diameter of 10 cm, which connected the moderator of the IBR reactor in Dubna on one side and a UCN detector on the other side. UCNs were stored in the neutron guide for 300 s, and such a long storage time allowed clearly separating UCNs from faster neutrons. Essentially, UCNs behaved in the guide very much like rare gas molecules.

Independently and simultaneously, in systematic measurements of total cross sections with very cold neutrons (VCNs) in Munich, extremely low neutron velocities were observed as a low-energy fraction of a broader velocity spectrum. Although the main property of UCNs, which consists of their storage in traps, was not demonstrated there, as low velocities as 5 m/s were indeed in the UCN energy range; they were measured using the time-of-flight method.

These breakthrough experiments were performed by the group of Fedor Shapiro[12] and by Albert Steyerl[13], respectively.

A major challenge of those experiments consisted of an extremely small fraction of UCNs in the initial spectrum of thermal neutrons in a moderator of a nuclear reactor. The UCN density achieved in these first experiments was of the order of $10^{-5} UCN/cm^{-3}$.

(*Note*: Even nowadays, when UCN densities of up to seven orders of magnitude higher than that became available for experimentalists due to using more intense initial sources of thermal neutrons and special superthermal sources of UCNs are being developed, the deficit of UCNs is still a major obstacle.)

---

[10]Zeldovich, Y. B. (1959). Storage of cold neutrons, *Sov. Phys. JETP* **9**, p. 1389.

[11]Vladimirsky, V. V. (1960). Magnetic mirrors, channels and traps for cold neutrons, *Sov. Phys. JETP* **12**, p. 740.

[12]Lushchikov, V. I., Pokotilo, Y. N., Strelkov, A. V., and Shapiro, F. L. (1969). Observation of ultracold neutrons, *JETP Lett.* **9**, p. 23.

[13]Steyerl, A. (1969). Measurements of total cross sections for very slow neutrons with velocities from 100 m/sec to 5 m/sec, *Phys. Lett. B* **29**, p. 33.

Exercise 3.10. *Estimate the fraction of UCN flux in the Maxwell spectrum of thermal neutrons with the temperature of T=300 K°.*

Note: Define here UCNs as those neutrons that could be trapped in a copper vessel.

After this short journey in neutron physics we have probably convinced the reader that the UCN is a promising candidate for observing a bouncing quantum particle. Its surprising property and the compulsory condition for feasibility of our experiments is its total and elastic reflection from material surfaces. Vladislav Luschikov and Alexander Frank noticed in 1978 such exceptional properties of UCNs for experiments with quantum bouncing on a surface[14].

An experiment aimed at the first experimental observation of neutron gravitational quantum states was performed at the Institut Max von Laue – Paul Langevin in Grenoble[15]. This experiment profited from such a particular property of a gravitational quantum state of a neutron as its macroscopic spatial extension; this property allows the use of a macroscopic experimental setup in order to scan directly the probability of observing the neutron.

Exercise 3.11. *Calculate:*

- The energy of three lowest gravitational quantum states of a neutron above an ideal reflecting mirror.

- The corresponding classical heights of raising a neutron in the gravitational field of the Earth.

- The mean values of neutron height for these three gravitational quantum levels.

- The mean momentum in each of these states.

---

[14]Luschikov, V. I., and Frank, A. I. (1978). Quantum effects occuring when ultracold neutrons are stored on a plane, *JETP Lett.* **28**, p. 1978.
[15]Nesvizhevsky, V. V., Boerner, H. G., Petukhov, A. K., Abele, H., Baessler, S., Ruess, F. J., Stoeferle, Th., Westphal, A., Gagarski, A. M., Petrov, G. A., and Strelkov, A. V. (2002). Quantum states of neutrons in the Earth's gravitational field, *Nature* **415**, p. 297.

From the above exercise one could conclude that a typical spatial size of low neutron gravitational states is tens of micrometers, i.e. a size which could be in principle resolved "with the naked eye". For instance, 10 microns is about standard thickness of the aluminum foil. The idea of the above experiment consisted of scanning the neutron density in gravitational states as a function of height, and to observe typical quantum features.

As all experiments with UCNs are heavily limited by the densities and fluxes available, we separated vertical and horizontal motions of UCNs, thus using a maximum range of horizontal velocities simultaneously and therefore improving statistics. To understand the idea of this experiment, imagine a flux of collimated neutrons with the mean total velocity of the order of 5–10 m/s and a small vertical angular divergency sent parallel to a very well-polished and flat glass plate. As long as the vertical motion of neutrons is considered, it is a superposition of many gravitational states.

Problem 3.3. *Using semi-classical (WKB) approximation, derive an expression for energies of gravitational quantum states of neutrons with large ordinal numbers $n \gg 1$.*

Estimate the accuracy of this approximation.

Verify the accuracy of this approximation even for the lowest quantum states, for which strictly speaking this approximation should be much less precise.

Estimate the number of neutron gravitational quantum states above a mirror that could be present in a divergent neutron flux with the velocity of 10 m/s.

A scheme of this experiment is shown in figure [3.1]. We would like to select a few lowest gravitational quantum states among many those states. For this purpose an *absorber*, i.e. a plate made of neutron absorbing material, is positioned above the bottom mirror at a given height, which could be adjusted within the range of values characteristic for gravitational states. (We are going to comment below on the principle of operation and properties of such absorbers.)

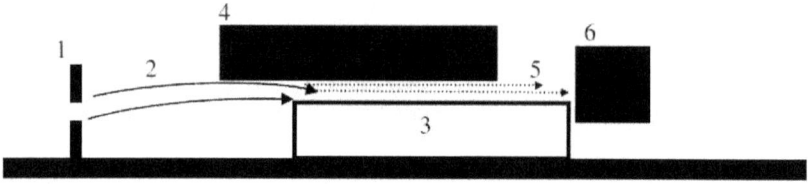

Fig. 3.1 A simplified scheme of measuring gravitational quantum states of neutrons. 1 – neutron collimator; 2 – illustration of classical trajectories of neutrons upstream entrance of installation; 3 – mirror; 4 – absorber/scatterer; 5 – illustration of horizontal component of neutron velocity in quantum regime of motion; 6 – neutron detector.

The role of the absorber is to selectively eliminate states with a vertical spatial size, which is larger than the size of the slit between the mirror and the absorber. A neutron wave, corresponding to such states, penetrates deep inside the absorber and thus is lost. Naively, one could think of thin-wall balloons, which are bouncing between a flat floor and a ceiling; the ceiling is starred with sharp nails. As soon as a balloon touches the ceiling it explodes; only balloons with a small enough height of bouncing survive.

Problem 3.4. *In the example with a flux of horizontally directed thin-wall balloons described above, assume that any contact of a balloon with the absorbing ceiling results in immediate explosion of the balloon; the horizontal velocity u is much larger than the spread in vertical velocities Δv, and the length of the mirror-absorber setup equals L. The angular distribution of velocities of the balloons is isotropic within the range Δv.*

Calculate the flux of such small classical elastic balloons, bouncing between the mirror and the absorber, at the exit of the slit as a function of the slit size H.

However, an analogous quantum picture would be much more interesting than that.

The most striking difference is that as soon as the size of the slit H between the mirror and the absorber becomes smaller than the characteristic spatial size of the lowest gravitational quantum state, no neutron could pass through this slit. This statement means that the flux of neutrons, measured at the exit of the slit, should *sharply decrease* as soon as the slit size becomes smaller than some critical value, which is close to the classical height of raising a neutron in the *first gravitational state* H_1.

Naively speaking, for the slit size smaller than the classical turning height H_1 corresponding to the first quantum state, one should expect that no neutron passes through the slit to the detector, like the prisoner who cannot squeeze through a hole in the fence in figure [3.2]. As soon as the slit size H is slightly larger than this value H_1 all neutrons settled in the ground gravitational quantum state pass through the slit to the detector with minor loss. In contrast, neutrons settled in excited quantum states should be totally absorbed.

Fig. 3.2   Escape from Alcatraz. The hole is an illusion.

As soon as the slit size starts to become equal to the classical height of raising neutrons settled in the second gravitational state, such neutrons would pass through with minor losses. But neutrons in the first gravitational state would also evidently pass through the slit without losses. Thus if one plots a flux of neutron at the exit of the slit as a function of the slit size, pronounced steps, corresponding to gravitational quantum states, should appear.

This picture differs significantly from the classical expectation, which imposes that the flux of neutrons in the detector is a monotonous function of the slit size.

Thus, the main idea of the first experiment on the observation of neutron gravitational quantum states is to identify *irregularities* (steps) in the neutron flux, measured in a detector at the exit of the described installation, as a function of the *slit size*. Such steps are the manifestation of a quantized vertical motion, i.e. allowed heights of neutron bouncing are a set of discrete numbers H_n. Each such number H_n should correspond to a predicted spatial size of a gravitational quantum state of a neutron.

In some sense this approach is equivalent to scanning the probability density to observe neutrons settled in quantum states. The scanning is performed using an absorber, which could be adjusted at various heights. The density probability is measured as a function of the height above the mirror. In the process of such scanning the absorber either eliminates neutrons settled at certain gravitational states (i.e. if $H < H_n$), or lets them pass through safely (i.e. if $H > H_n$).

Let us discuss briefly the *experimental feasibility* to realize this approach. A crucial point concerning the feasibility is that the mentioned characteristic heights H_n have macroscopic values. The reason to underline this point is twofold.

First, the mirror and absorber plates should deviate from *parallelism* to a minimum extent, otherwise we could not establish accurately the absorber height H (one could say that the relative inclination of the mirror and the absorber contributes into the experimental accuracy of measuring the slit size H), and also the interaction of a UCN with a slit with a size evolving in time (in the frame of reference related to the horizontally moving UCN) might induce parasitic transitions between quantum states.

What is the *optimum length* of the mirror and absorber setup? This question turns to have a much deeper sense than just a technical issue, and thus we are going to discuss it in more detail below. Concerning the technical feasibility, the length of installation used in the first experiment was equal to 10 cm. It is practically possible to install a mirror and an absorber of such length parallel to each other with the absolute accuracy of 1 micron and to control the slit size between them with the same accuracy.

Second, an experiment could provide a meaningful result only if the total number of neutrons detected at the exit of the installation for each value of the slit size H is larger than the number of *background events*, and also if the total number of detected neutrons would allow a reasonable statistical accuracy in each point. In fact, with unique extra-low-background detectors developed especially for these quantum gravitational experiments, those two conditions coincide.

Namely, as far as the number of background events per point (per a value of the slit size H) becomes comparable to one, its further decrease would not be very useful; it would not improve considerably the resulting accuracy and sensitivity of the experiment. As a maximum total measuring time is typically equal to a few days, and thus a maximum measuring time per point is equal to a few hours, the critical value of a detector background is equal to the reciprocal measuring time per point $10^{-4}s^{-1}$.

That low background level was indeed achieved. The residual detector background was defined by alfa-activity of internal walls of the detector, by unavoidable presence of fast neutrons in the reactor hall (the detector could be very efficiently protected against external thermal neutrons, therefore they do not contribute significantly), by UCNs scattered somewhere inside the vacuum chamber of the experimental setup and penetrating around the mirror-absorber assembly to the detector, by cosmic background, and so on.

If the background is negligible, the number of events N determines the *statistical accuracy* of measured results. A statistical uncertainty decreases as $1/\sqrt{N}$ with accumulating statistics. The reader already knows that the UCN is a small fraction in the spectrum of a nuclear reactor, thus that the UCN flux is relatively small. However, the problem is that only a small fraction of this small fraction could be used. Obviously the UCN flux changes as a function of the value of the slit size H.

Problem 3.5. *A point-like source of neutrons is positioned 1 m below and 1 m upstream from the entrance to an experimental setup consisting of a mirror and an (parallel to it) absorber installed above the mirror; their sizes are 10 by 10 cm. The neutron velocity modulus is distributed according to Maxwellian law with the characteristic temperature of $T=10$ mK°; the angular distribution of neutrons is isotropic. The size of the slit between the mirror and the absorber is equal to the turning height of the lowest gravitational quantum state $H = H_1$. The mirror totally reflects all neutrons and* *the absorber eliminates all neutrons, which could raise to its height.*

Calculate, using the classical approximation, the fraction of neutrons which could pass through the slit.

A way out in such a kind of low-count-rate experiment is to *accumulate statistics* as long as they are feasible. Although this way seems to be evident, it is constrained by the large cost of beam time at a high-flux reactor. The detector count rate in the actual experiment was of the order of 10^{-2} s^{-1} in the regime of measuring the lowest quantum states. Therefore, one has to perform a measurement for hours, continuously for every slit size in order to achieve reasonable statistics.

Another difficulty of this experiment is that the component of neutron velocity parallel to the mirror surface is larger than a typical vertical component of velocity of neutrons in gravitational quantum states by three orders of magnitude. This estimation means that even a minor *waviness of the mirror* would mix (large) horizontal and (small) vertical velocity components. In such a case the neutron would raise high enough to be absorbed, and the gravitational state would be destroyed.

This disturbing effect imposes a very tight constraint on the quality of a mirror, which is needed to provide that most reflections are highly specular. In particular, the deviations from the *flatness* of the mirror should be at least smaller than 1 micrometre, and the roughness of the mirror surface should be smaller than at least 1 nanometer. Such parameters are sufficient to observe gravitational states. Below we are going to study how they affect lifetimes of gravitational quantum states.

The same kind of problem is produced by even tiny *vibrations* of the installation, particularly those with frequencies comparable to the frequencies of transitions between different gravitational quantum states. Such vibrations would result in a vertical kick that adds some extra vertical velocity to the neutron settled initially in a gravitation state. To avoid such parasitic transitions, dangerous vibrations are suppressed by the installation design and by a nontrivial anti-vibrational shielding.

To gain better understanding of the methodical and experimental tricks used, the reader could look at review articles[16,17], which also contain references to principle original publications.

[16] Baessler, S. (2009). Gravitationally bound quantum states of ultracold neutrons and their applications, *J. Phys. G* **36**, p. 104005.
[17] Nesvizhevsky, V. V. (2010). Near-surface quantum states of neutrons in the gravitational and centrifugal potentials, *Physics-Uspekhi* **53**, p. 645.

Problem 3.6. *Consider, in the classical approximation, UCNs passing through the ideal assembly of a horizontal flat mirror on bottom and a rough absorber on top ("ideal" means here that the assembly is long enough to provide that each UCN in the slit between the mirror and the absorber would touch the mirror and the absorber [if the vertical energy allows] at least once; UCN losses in the mirror are negligible; each UCN is totally lost as soon as it touches the absorber). Imagine that quantum states of UCNs do not exist, the motion of UCNs is not governed by quantum mechanics, and instead it is described purely by classical mechanics at all scales. Imagine that UCNs are distributed uniformly in the phase-space at the entrance to the slit.*

Calculate analytically, in the classical approximation, the flux of UCNs through the slit as a function of the slit size.

A *characteristic result* of this experiment is shown in figure [3.3]. One can clearly observe a range of slit sizes, which correspond to the absence of any flux penetrating through the slit. This observation is possible due to very low level of background in a specially designed and shielded neutron detector. One can also see that irregularities, which correspond to the two lowest gravitational quantum states, are established within the experimental accuracy. However, they do not look sharp, as we would expect from our preceding simplified discussion. They are practically washed out for higher states $n > 2$.

Exercise 3.12. *Use the result for the classical UCN flux through the ideal slit described in the previous problem and trace it in figure [3.3] as a function of the slit size.*

Why are all gravitational states with high quantum numbers so poorly resolved? Could we detect them? Is there a principal phenomenon, which limits the spectrometer resolution? Can the present resolution be improved

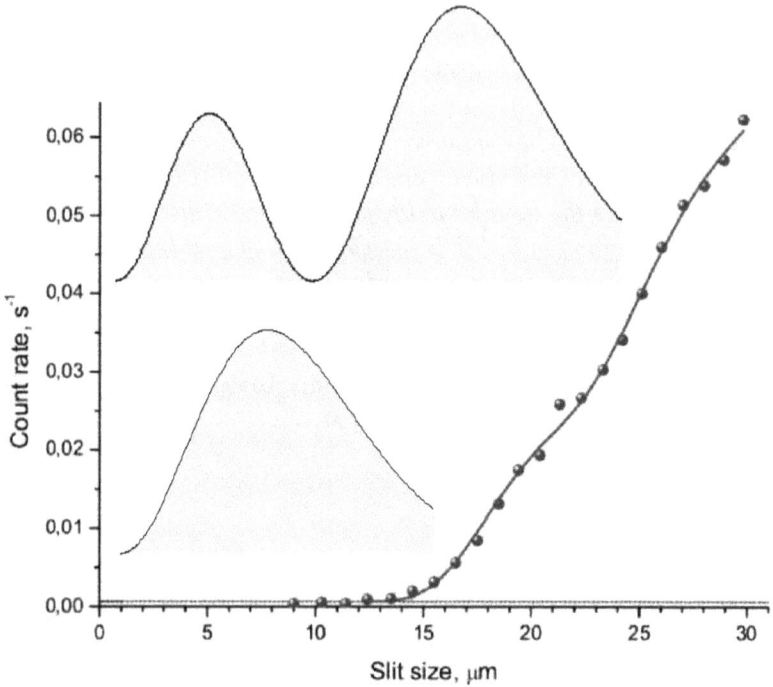

Fig. 3.3 A typical result of first experiments on searching for gravitational quantum states of UCNs. The measured neutron count rate in the detector is shown with circles as a function of the size of slit between a flat polished horizontal mirror on bottom and an absorber with microscopically rough and macroscopically flat surface on top. A theoretical curve fits the data. Background is indicated with a solid horizontal line. One clearly observes the main feature of the measured data, which consists of the fact that UCNs do not pass through the slit as long as the slit size is smaller than the characteristic size of the lowest gravitational quantum state. One also observes irregularities corresponding to second and third quantum states. Predicted probability density is shown in the insert for first and second quantum states.

due to improving extensively certain parameters of the present experiment? What accuracy could we achieve while studying neutron gravitational states? Why do we want that? What could we learn from precision measurements?

These are the questions which we are going to discuss soon.

3.2 Bouncing Wave: Penetration Through a Wall and Reflection from a Well

Observation of an individual gravitational quantum state of neutrons is rather challenging, as the reader has probably learned from our preceding discussion. The mentioned difficulties, which are mainly related to the small amount of UCNs that we have at hand as well as to the ultimate quality of mirrors (at the level of highest optical standards) that we have to use are compensated to some extent by major advantages: the electrical neutrality and small magnetic moment of the neutron, plus its small mass.

The latter properties of neutrons allow us to avoid, or at least to suppress, false effects originating from electromagnetic forces as well as to measure the spatial density distribution of neutrons in the lowest gravitational states using macroscopic devices. In an ideally designed experiment, what would be the principal limitations of such measurements in a case if we limit ourselves only by the method of scanning the spatial distribution of neutron density?

If we are not satisfied with the accuracy that could be achieved in experiments that measure the spatial distribution of neutron density by means of scanning the density with an absorber or that measure it directly using a position-sensitive detector of sufficiently high spatial resolution (better than 1 micrometer in the vertical direction), could we choose some other observables in order to get more precise information about gravitational states?

Looking at figure [3.3], the reader could be disappointed with the fact that irregularities (corresponding to quantum features) in the curve, which describe the neutron flux as a function of the slit size, do not look very pronounced, while the statistical accuracy of the data is quite convincing in each point. In order to understand the reason for such a modest spatial resolution of the experiment let us clarify the effect and the efficiency of the absorber.

As we have discussed above, an ideal absorber should totally eliminate all neutrons settled in a gravitational quantum state if the absorber height H is smaller than the effective size of the corresponding gravitational state H_n (the turning height for a classical particle with equivalent energy), and it should leave unaffected all neutrons settled in such quantum states that the condition $H > H_n$ is valid. However, an absorber with such properties does not exist.

The reason for the latter concern is in the already-mentioned *tunneling effect*; tunneling through (under) the gravitational potential in this particular case.

In order to reveal the sense of this statement, let us analyze what exactly happens if the absorber is adjusted slightly above the classical height of the lowest gravitational state H_1. As the reader has already learnt, the probability density to observe neutrons settled in a gravitational quantum state is nonzero everywhere above the classical turning height H_1. One could check that the probability to observe a neutron settled in the lowest gravitational state above the height H_1 is about 14%.

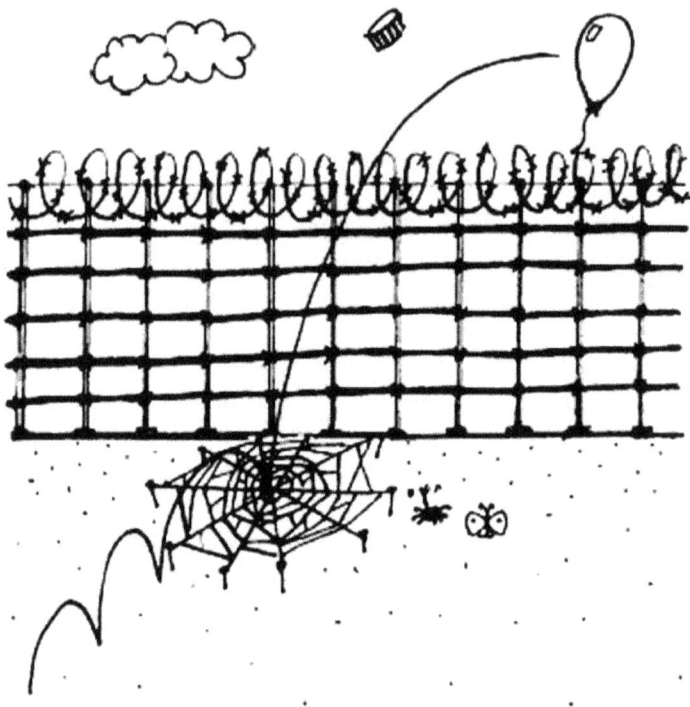

Fig. 3.4   Escape from Alcatraz. The wall is an illusion.

Such an effect would allow a prisoner to overcome a jail wall, as shown in figure [3.4]. In quantum world, an object would be observed sooner or later in the classically prohibited region.

Exercise 3.13. *Calculate the probability to observe a neutron settled in the lowest gravitational state above the classical turning height H_1, i.e. in the classically forbidden zone.*

This estimation means that a neutron wave "feels" an effect of an absorber even if the absorber is installed above the critical height H_1. Moreover, the probability to observe a neutron *higher than the critical height* decreases smoothly while the absorber height increases. Thus the neutron flux in the detector does not look like a step function of the slit size, it increases *smoothly* because neutrons are penetrating through the gravitational barrier inside the absorber.

It is important to mention that the presence of an absorber at a certain height above the mirror not only defines the probability of neutron tunneling through the gravitational barrier into the absorber and thus the decay of a quantum state, but it also modifies the gravitational state itself. In particular the energy of a gravitational state changes. In order to derive a qualitative relation we need to have a realistic model of behavior of a neutron wave inside an absorber.

A useful approach for solving this problem is to model an absorber as a medium with a complex optical potential. The real part of this potential is negative and corresponds to an attractive potential; its imaginary part is also negative and describes absorption:

$$U_{abs}(z) = (-U_0 - iV)\,\Theta(z - H) = \begin{cases} -U_0 - iV, & z \geq H, \\ 0, & z < H. \end{cases} \qquad (3.6)$$

The Hamiltonian, which describes neutron motion in the presence of an absorber, looks like:

$$\hat{H} = \left(-\frac{\hbar^2}{2m}\frac{d^2}{dz^2} + Mgz + U_{abs}(z) \right). \qquad (3.7)$$

The value of energy $\widetilde{E}_1$ in the gravitational quantum state perturbed by an absorber is a mean value of the above-mentioned Hamiltonian (3.7). Taking into account that the perturbation by an absorber is small provided that the absorber is positioned above the classical turning height (soon we are going to get a qualitative estimation of the degree of such smallness), we could get the following expression for the gravitational state energy in the presence of an absorber:

$$\tilde{E}_1 \simeq E_1 - (U_0 + iV) \int_H^\infty |\Psi_1(z)|^2 dz. \tag{3.8}$$

Here E_1 is the unperturbed value of the gravitational state energy in the absence of an absorber.

The integral on the right-hand side of the equation (3.8) is small, because the integration is performed over the heights, where the neutron wave decays rapidly (as a function of height) inside the classically forbidden region. The probability of decay is proportional to the wave-function modulus square, measured at the absorber height H. This integral calculates a complex shift of the gravitational state energy due to the presence of a complex absorber optical potential (3.6).

The imaginary part of the complex energy value (3.8), which is also called a half-width $\Gamma/2$ of the quantum state, is related to a characteristic time of decay of the neutron quantum state (its lifetime) as $\tau = \hbar/\Gamma$. Thus the state width is proportional to the probability density to observe a neutron at the absorber height, or the so-called tunneling probability:

$$\frac{\Gamma}{2} \sim V_0 |\Psi_1(H)|^2. \tag{3.9}$$

Exercise 3.14. *Using the asymptotic expression for the Airy function of a large positive argument* $\mathrm{Ai}(x) \approx \frac{1}{\sqrt[4]{x-\lambda}} \exp\left(-2/3(x-\lambda)^{3/2}\right)$ *(2.42), verify that the expression for the probability density to observe a particle under the gravitational barrier at a height $H > H_1$ is the following:*

$$P \approx \sqrt{\frac{H - H_1}{H_1}} \exp\left(-4/3(\frac{H - H_1}{l_0})^{3/2}\right). \tag{3.10}$$

Here l_0 is the characteristic gravitational quantum scale $l_0 = \sqrt[3]{\frac{\hbar^2}{2mMg}}$ (2.38).

Problem 3.7. *A neutron wave is characterized by a given energy E in the gravitational field with the strength g. A horizontal plate, which is characterized by an attractive optical potential $-U_0$, is installed at a height H.*

Calculate the ratio of the neutron flux inside the medium to the incoming neutron flux.

Derive the exact result as well as the result in semi-classical approximation.

Compare them with each other and also with the result of the previous exercise.

Thus we see that a neutron wave enters into an absorber with the amplitude, which is determined by the tunneling through the gravitational barrier. This is the essence of the tunneling effect.

What happens to a neutron wave inside an absorber? Typical values of the absorber optical potential U_{abs} are many orders of magnitude larger (10^{-8} eV) than the gravitational state energies (10^{-12} eV), thus we could neglect gravity and the neutron wave-function inside the absorber could be considered as a plane wave:

$$\Psi_{abs}(z) \sim \exp\left(i\sqrt{2m(U_0 + iV)}z\right). \tag{3.11}$$

Due to the presence of the imaginary part of optical potential, which describes the physical effect of absorption, the outgoing wave decays inside the bulk of the absorber. Thus the part of the neutron wave, which penetrates through the gravitational barrier into the absorber, propagates into the bulk of the absorbing medium and it disappears. One could see that the rate of absorption, given in the imaginary part of the absorber's optical potential, could be very small; the important message is that no neutron returns back.

In other words the neutron wave inside the absorber is a purely outgoing wave. The effect of the absorber is thus equivalent to the boundary condition, which would be applied to the wave-function at the absorber height H and which consists of the statement that the neutron wave at the heights larger than the absorber height $z > H$ is purely outgoing.

Now we are prepared to combine all these results together and to obtain a qualitative result for the *lifetime* of a neutron gravitational quantum state in the above-described experiment.

Problem 3.8. *Derive an equation for the complex energies of gravitational quantum states in the presence of an absorber, which is described with an optical potential* $-(U_0 + iV)\Theta(z - H)$ *with the imaginary part in the range* $E_n \ll V \ll U_0$, *which is installed above the classical turning height* $H > H_n$.

Using the asymptotic expression for the Airy functions $\mathrm{Ai}(x) \approx \frac{1}{\sqrt[4]{x-\lambda}} \exp\left(-2/3(x-\lambda)^{3/2}\right)$ *(2.42), verify that the width* Γ_n *of the gravitational state could be approximated via the following expression:*

$$\frac{\Gamma_n}{2} \simeq \frac{\hbar M g}{\sqrt{2mU_0}} \sqrt{\frac{H - H_n}{H_n}} \exp\left(-4/3(\frac{H - H_1}{l_0})^{3/2}\right). \qquad (3.12)$$

This very important expression (3.12) explains qualitatively the origin of the observed unavoidable smoothing of quantum steps in the experimentally observed dependence of the neutron flux through a slit as a function of the slit size.

The wave-function $\exp(-iEt/\hbar)$ of a gravitational quantum state of a neutron evolves as a function of time in such a way that it contains a decaying factor, which is parameterized via the imaginary term of the energy, $exp(-\Gamma t/(2\hbar))$. If a neutron spends a certain time τ_{pass} in the slit between the mirror and the absorber then a given gravitational state is suppressed by the absorber (we assume that $H > H_n$), and the transmission factor is the following:

$$F = \exp\left[-(\Gamma \tau_{pass})/\hbar\right]. \qquad (3.13)$$

The transmission factor F increases rapidly with increasing the size H of the slit between the absorber and the mirror as long as the transmission is smaller than unity, then it evidently has to saturate. This relatively rapid, although not immediate, increase of the transmission factor is governed by the exponential decrease of the tunneling probability, and thus the exponential decrease of the width of the quantum state Γ, as clear from expression (3.12).

One could also conclude from this expression that the *uncertainty in the position of a "step"* (a smoothing of the quantum "step") is of the order of the value of the gravitational quantum length scale l_0. (Further on we are going to derive a more accurate expression for this uncertainty.)

At this point we face another difficulty, namely that concerning *too large density* of the spectrum of gravitational states. The problem is that with increasing a quantum number the energy levels become more and more dense, and thus the classical turning heights H_n become more and more dense as well. Therefore, as soon as the distance between "steps" becomes smaller than the uncertainty in the step position, it is not possible to resolve the neighboring states in the neutron flux.

Problem 3.9. *Verify that the difference between classical turning heights H_n of gravitational quantum states with quantum number n decreases as follows $n^{-1/3}$.*

Estimate the lowest quantum number n of a gravitational quantum state, for which the following condition is valid: $H_{n+1} - H_n \leq l_0$.

The reader could notice that the absolute spacing between classical turning heights, which correspond to neighboring gravitational quantum states, is of the order of the characteristic gravitational quantum state l_0 even in the best case of the lowest states. We have just established that an uncertainty in the "step" height in the neutron flux is also of the order of the characteristic gravitational quantum state l_0 due to the existence of the tunneling effect.

Thus, only a few lowest gravitational quantum states could be resolved by means of the method of scanning the neutron spatial density using an absorber in the experimental configuration described above. And the reason for this constraint is purely quantum. It consists of two factors: the penetration of a neutron wave through the gravitational potential barrier on one hand, and the increasing of the level density with increasing a quantum number on the other hand.

To what extent could we improve our ability to resolve neighboring quantum states, without modifying the principle of observation of such quantum states? How do the concrete properties of the absorber enter into the game, and to what extent? What is the role of time, spent by a neutron in the slit between the mirror and the absorber?

To shed light on these questions, let us study more accurately how the width of a gravitational state Γ changes as a function of the absorber properties and the time of passage τ_{pass}.

For convenience, we introduce a characteristic "time of absorbtion" τ_{abs}, defined as follows:

$$1/\tau_{abs} = \frac{Mg}{\sqrt{2mU_0}} \sqrt{\frac{H - H_n}{H_n}}. \tag{3.14}$$

Then we can write down expression (3.13) in a different form:

$$\ln F = \exp\left(-4/3(\frac{H - H_1}{l_0})^{3/2}\right) \frac{\tau_{pass}}{\tau_{abs}}. \tag{3.15}$$

Problem 3.10. *Verify that the uncertainty δ in the height of a "step" in the transmittance as a function of height in the case of sufficiently long observation (passage) times $\ln(\tau_{pass}/\tau_{abs}) \gg 1$ is equal to:*

$$\delta \sim \frac{l_0}{(\ln(\tau_{pass}/\tau_{abs}))^{1/3}}. \tag{3.16}$$

Problem 3.11. *Verify that the shift Δ of a "step" in the neutron transmittance as a function of height relative to the value of a corresponding classical turning height H_n in the case of sufficiently long observation (passage) times $\ln(\tau_{pass}/\tau_{abs}) \gg 1$ is equal to:*

$$\Delta \sim l_0(\ln(\tau_{pass}/\tau_{abs}))^{2/3}. \tag{3.17}$$

The above results show that we could decrease the uncertainty in the position of a measured "step" only at the expense of a very significant increase in the ratio of the passage time and the absorption time τ_{pass}/τ_{abs}. Indeed, a factor $(\ln(\tau_{pass}/\tau_{abs}))^{1/3}$ is a *very slow function* of these two parameters. At the same time a position of such a "step" would be shifted relative to the position of a corresponding classical turning height H_n, which also depends only slightly on the ratio τ_{pass}/τ_{abs}.

An increase of the ratio τ_{pass}/τ_{abs} by means of increasing the time of passage is *severely limited* by two factors. On one hand, at least at the present level of technology, it is extremely difficult to produce a parallel system of a mirror and an absorber of high enough quality with the length of more than a few dozens of cm; on the other hand, a maximum available or projected phase-space density of UCNs does not allow us to get reasonable statistics with UCNs slower than a few meters per second.

However, the decrease of the absorption time τ_{abs} is *also limited*, and this limitation is of principle character. Indeed, using a classical analogy, we could estimate quantitatively this constraint and state that the time of absorption of a neutron, which is bouncing with a given energy above a mirror, is equal in the very best case to the time of flight between the mirror and the absorber. It corresponds to the intuitively transparent semiclassical expression for the lifetime τ_{WKB}:

$$1/\tau_{WKB} \approx \omega_n P. \tag{3.18}$$

Here ω_n is a classical oscillation (bouncing) frequency; P is a probability to absorb a neutron in one collision with an absorber. Even in the case $P = 1$, the classical lifetime is defined as the time of flight between the mirror and the absorber. Could we increase the imaginary part or the real part of the absorber optical potential? Would this effort improve the absorbing properties, and would it increase the absorbing probability?

In order to clarify these points let us look more attentively at expression (3.14), which describes the effect of an absorber. A surprising fact, which is following immediately from this expression, is that the absorbing properties are inversely proportional to the momentum of a neutron wave inside the absorbing medium. In other words, we face a paradox: the deeper the absorber potential, the weaker the absorbing power of such an absorber!

In order to understand this counterintuitive conclusion we would like to remind the reader that the ideal absorber is one that which does not reflect a neutron wave back. Thus the reduction of absorbing properties could be related to the reflection of a neutron wave from a deep optical potential of the absorber. This hypothesis looks strange from the classical point of view. In order to verify it we propose that the reader solves the following problem.

Problem 3.12. *Calculate the amplitude S of reflection and the probability of reflection of a wave, which is propagating from the right to the left with the energy E and interacts with a deep optical potential $U_{abs} = (-U_0 - iV)\Theta(-x)$. Consider the case of relatively small energies $E \ll |-U_0 - iV|$.*

Calculate the scattering length, which is determined as follows: $a = (1 - S)/(2ik)$, where $k = \sqrt{2mE}$.

Note: Use the continuous condition at the interface between free space and the absorber at $x = 0$.

Solution of the above problem provides a key for resolving the above-mentioned paradox. Indeed, the amplitude of reflection of a wave from a *deep well* $(E \ll |-U_0 - iV|$ turns out to be:

$$S \simeq 1 - 2i\frac{\sqrt{2mE}}{\sqrt{2m(E + U_0 + iV)}} = 1 - 2i\frac{\lambda_{abs}}{\lambda_{in}}. \tag{3.19}$$

Here $\lambda_{abs} = \frac{\hbar}{\sqrt{2m(E+U_0+iV)}}$ is an effective neutron wavelength inside the absorbing medium, and $\lambda_{in} = \frac{\hbar}{\sqrt{2m(E}}$ is a wavelength of the incoming neutron wave.

As we see, the deeper the potential well, the closer the reflection amplitude to the unit (i.e. to the maximum possible value). Thus, we confirm with a simplified example that the physical reason for strong reflection of a neutron wave from an absorber is a smallness of the ratio of the energy of the incoming wave to the potential depth. In order to achieve a better understanding let us clarify the role of the *steepness* of a potential. The following problem is helpful.

Problem 3.13. *Calculate the amplitude S of reflection and the probability of reflection of a wave, which is propagating from the right to the left with the energy E and interacts with a deep optical potential parameterized in the following form:* $U_{abs} = (-U_0 - iV)\frac{1}{\exp(x/a)+1}$.

Consider the case of relatively small energies $E \ll |-U_0 - iV|$ *and relatively smooth potential* $\frac{\hbar}{\sqrt{2mE}} \ll a$.

Note: Use the WKB approximation.

Solution of this problem consists of an important conclusion that in the case of a slowly changing effective neutron wavelength, $\hbar/\sqrt{2m(E - U_{abs}(x))}$, the WKB approximation is valid everywhere, and the wave-function keeps its form of the incoming wave, $\exp(-i\frac{1}{\hbar}\int \sqrt{2m(E - U_{abs}(x))}dx)$ everywhere. Thus, no reflected wave appears and the amplitude S tends to zero. This is a case of an *ideal absorber*! An absorber is perfect in the case of its optical potential changing slowly enough.

An effect of reflection of a wave from an attractive potential is related to the steepness of this potential. In the region, where the potential changes faster than the effective wavelength does, the reflected wave is generated. This phenomenon is of purely quantum nature; it is called *quantum reflection*. Below we are going to study this phenomenon in the case of antiatoms; the quantum reflection allows one to keep antiatoms bouncing above a material surface.

However, if an efficient absorption is required then the quantum reflection is a drawback.

Now the reader could easily understand the physical limitation on the efficiency of an absorber. A typical depth of the optical potential of a close-to-ideal neutron-absorbing medium, like polyethylene, is of the order of 10^{-8} eV, while a typical energy of lowest gravitational states is of the order of 10^{-12} eV. This comparison prompts that such an absorber with a well-defined flat surface would reflect a significant part of the incoming wave.

And there are no other realistic parameters of the problem that could affect our conclusion.

The method we use for enhancing the absorbtion efficiency is to make the interface between vacuum and an absorbing medium not steep but *diffuse*[18], i.e. the optical potential should increase inside the medium from zero to its maximum value within a range of the order of the typical gravitational wavelength l_0. In this case the neutron wavelength decreases adiabatically with increasing the optical potential, and thus practically no reflection takes place.

This gain in the absorber efficiency is achieved at the expense of the fact that an absolute height of the absorber could not be defined better than within an uncertainty equal to l_0.

Problem 3.14. *Calculate the value of optical potential U_p of polyethylene using neutron scattering data and physical properties data.*

Calculate the probability of reflection of a neutron with the energy equal to 10^{-12} eV from such an optical potential.

Calculate the width of the first gravitational state as a function of the absorber height for above-critical heights $H > H_1$.

Problem 3.15. *Calculate the probability of reflection of a neutron with the energy equal to 10^{-12} eV interacting with an absorber with a diffuse optical potential in the form $U = U_p \frac{1}{\exp(-x/a)+1}$ with the characteristic range equal to the gravitational quantum scale $a = l_0$.*

Calculate the width of the first gravitational state as a function of the absorber height for above-critical heights $H > H_1$.

Problem 3.16. *Verify that the expression $1/\tau_{abs} = \frac{Mg}{\sqrt{2mU_0}}\sqrt{\frac{H-H_n}{H_n}}$ (3.14) could be rewritten as follows:*

$$1/\tau_{abs} = \frac{|\lambda_{abs}|}{|\lambda_{grav}|}\omega_n, \qquad (3.20)$$

---

[18]Nesvizhevsky, V. V., Boerner, H. G., Gagarski, A. M., Petrov, G. A., Petukhov, A. K., Abele, H., Baessler, S., Stoeferle, Th., and Soloviev, S. M. (2000). Search for quantum states of the neutron, *Nucl. Instrum. Meth. A* **440**, p. 754.

where λ_{abs} is an effective neutron wave length inside the absorber, λ_{grav} is an effective neutron wavelength at the height H under the gravitational barrier, and ω_n is a classical bouncing frequency of a neutron with the energy E_n.

Now it becomes clear that effective absorbing of neutrons in lowest gravitational states is a rather nontrivial task. The practical solution for enhancing the absorbtion consists in using a scatterer with *rough surface* instead of an absorber with a plane surface. A rough surface results in non-specular reflections of a neutron, which is thus *mixing* a large horizontal velocity and a small vertical velocity of the neutron. The roughness amplitude of such a scatterer is about a few micrometers.

A rough scatterer is equivalent to a plane absorber with a smooth interface between the vacuum and the absorbing medium with the range of diffuseness at the interface equal to a few micrometers. Such an absorber-scatterer is much more efficient than a simple absorber, and now we understand why.[19]

Let us mention another problem related to any precise absolute measurements with an absorber.

As one could conclude from expression (3.17) a large value of the ratio of the passage time and the absorption time τ_{pass}/τ_{abs}, which is required for improving the resolution (3.16) of a "step" height, results in a *shift of the step*. Such a shift, which is a function of the absorber's properties, limits any high-precision studies with the described experimental setup. The reason for this uncertainty is that the properties of an absorber, which enter into the measured height of a "step", could be controlled to a finite extent.

Thus, the observed effect of the smoothing of quantum "steps" in the neutron flux as a function of the absorber height above the mirror, which limits the experimental spatial resolution, is not a consequence of parasitic experimental uncertainties; it has an important physical background. Surprisingly, the quantum effect of *tunneling* through the gravitational barrier and the quantum effect of *reflection* from a steep potential, mask the phenomenon of *quantizing* gravitational energy levels.

---

[19] Escobas, M., and Meyerovich, A. E. (2011). Beams of gravitationally bound ultracold neutrons in rough waveguides, *Phys. Rev. A* **83**, p. 033618.

Let us underline that an absorber is used in the described experiments for scanning the neutron density as a function of height, while a detector performs the integral counting of all neutrons that pass through the mirror–absorber setup. An alternative option consists of studying the spatial characteristics of gravitational states by means of "making a direct snapshot" of a profile of the spatial density distribution using a *position sensitive neutron detector.*

Such a detector, like a photo-sensitive film in the case of photography, keeps a trace resulting from the event of a neutron hit in the converter at the detector surface. The detector has to provide a resolution of a micrometer or better in order to resolve quantum structures in the probability distribution; detectors with such extremely high spatial resolution were developed especially for experiments with gravitational quantum states of neutrons[20]. A real-scale picture of a neutron density could be (and has been) achieved in such a way.

Problem 3.17. *Verify that for a broad distribution of horizontal velocities of UCNs the spatial density distribution over height, snap-shorted in a position-sensitive detector, is:*

$$P = \sum_{i=1}^{N} \frac{p_i}{|\operatorname{Ai}'(-z_i)|^2} |\operatorname{Ai}(z)|^2, \qquad (3.21)$$

where z_i is i−th zero of the Airy function and p_i is a population of i−th gravitational state.

Consider that the absorber height H is larger than the n-th turning height $H \geq H_n$.

Calculate the heights of maxima and minima of the above distribution for third quantum state $n = 3$.

Problem 3.18. *Derive the relation between zeros of gravitational quantum states with different quantum numbers n.*

──────────────────────

[20]Baessler, S., Gagarski, A. M., Lychagin, E. V., Mietke, A., Muzychka, A. Yu., Nesvizhevsky, V. V., Pignol, G., Strelkov, A. V., Toperverg, B. P., and Zhernenkov, K. (2011). New methodical developments for GRANIT, *Compt. Rend. Phys.* **12**, p. 729.

If a few gravitational quantum states are selected using an absorber, the measured neutron density distribution (3.21) would manifest a set of maxima and minima, which reflect the spatial quantum properties of selected gravitational states. Indeed, each of the squares of the Airy functions, which enters in the expression for the neutron density in the position-sensitive detector (3.21), has some pronounced maxima and minima.

The method of position-sensitive detectors would be beneficial if it provided a better resolution of these maxima and minima.

Here is a good place to remind the reader about the tunneling effect. This effect results in the overlapping of the "tails" of lower gravitational states with minima of higher excited states. This overlapping of neighboring states results in the "washing out" of maxima; they become less pronounced. Another obstacle in developing this method is a low intensity of neutron flux at the detector, thus long expositions are required to get a "good snapshot" of a neutron-wave density distribution.

Could we propose a method not only to expose clearly the quantum properties of gravitational states, but also to measure these properties with the highest possible accuracy?

What physical observable is the most appropriate for such studies?

3.3 Energy of Gravitational States: Shaken, Not Stirred

A peculiar feature of neutron gravitational quantum states, which consists of their macroscopically large spatial extension, makes it possible to prove their existence and directly measure a distribution of the spatial density of neutrons in such quantum states, in particular using several experimental methods described in the previous sections. However, serious limitations constrain any precision measurements of the spatial properties of gravitational states.

A natural question arises: is there another physical observable that could be more promising for precision experimental studies of gravitational states? Experimentalists know that a physical value, which could be measured with the highest accuracy, is *frequency*; measurements of frequency profit from the periodic motion. In the quantum world, frequency is related to energy via the famous Planck constant $E = \hbar\omega$; and classical periodic motion corresponds to quantum interference.

The world accuracy records are associated with measurements of the frequency of transitions between energy levels in hydrogen atoms. Could a neutron bouncing in gravitational quantum states above a mirror, become a "gravitational analog" of the hydrogen atom?

A crucial underlying idea in such a kind of measurement consists of using the *resonance* phenomenon. In general terms, resonance means the coincidence of a frequency of some external periodic perturbation with an internal characteristic frequency of a system. Such a coincidence manifests itself differently in classical systems than in quantum systems, although one manifestation could be reduced to another by a proper correspondence of parameters of such resonance.

In classical physics, a resonance is usually associated with a dramatic continuous increase of the energy, or the amplitude, of oscillations of a system. The quantum system is characterized by a set of quantum states. Under the influence of a periodic external force the quantum system demonstrates intensive transitions between initial and final quantum states as far as the frequency of an external periodic field coincides with a transition frequency between initial and final states $\omega_{nk} = (E_n - E_k)/\hbar$.

To clarify this issue let us find out what happens with a quantum system under the influence of a temporally changing force.

It is instructive to start our analysis with a *perturbation* theory approach.

Problem 3.19. *Verify that the probability to observe a bouncing neutron in n-th gravitational state $\psi_n(z)$ above a horizontal mirror under the influence of a weak time-varying potential $U(z,t)$, provided that initially the neutron was settled in i-th gravitational state $\psi_i(z)$, could be found from the following expression:*

$$P_n(t) = |c_n(t)|^2, \qquad (3.22)$$

$$c_n(t) \simeq -\frac{i}{\hbar} \int_0^t U_{in}(\tau) \exp\left(-i(E_n - E_i)\tau/\hbar\right) d\tau, \qquad (3.23)$$

$$U_{in}(t) = \int_0^\infty \psi_i(z)\psi_n(z)U(z,t)dz. \qquad (3.24)$$

Note: Assume that all amplitudes c_k except for c_i are small, and also neglect "backward" transitions into the initial state.

Estimate the limits of validity of the above results in the case the perturbation $U(z,t)$ has the following form $U(t,z) = U_0 \frac{z}{l_0} \Theta(t - t_0)$.

A very important message, which could be derived after analyzing the above expressions (3.22, 3.23), is that an initially prepared stationary state turns into a *nonstationary* state as soon as it starts to be affected by a time-dependent external force. Moreover, an effect of such an external time-dependent influence is that it would result in a relatively small, but nonzero, probability $P_n(t)$ to measure a *however large* energy of a bouncing neutron.

This result is known as the Dirac perturbation theory. The predictive power of the perturbation approach is limited by those cases, in which the effect of an external force is relatively small. In particular it is no longer possible to use it as soon as the calculated transition probabilities would become close to unity; for instance, backward transitions would considerably affect the behavior of the system in this case. However, this method provides us with a closed-form result for any kind of small perturbation.

An interesting consequence of this simple treatment, outside the quantum bouncing problem, is that we have to revisit our explanation of the stability of quantum systems such as atoms.

Indeed, atoms (as well as any other quantum system) never exist in complete isolation from the external *environment*, even if special efforts are devoted to suppress the external perturbations. Such an environment could consist of various fields of other quantum particles, atoms, etc. If a system is properly identified (for instance, a tightly bound quantum system like an atom), usually its interaction with the environment looks like the interaction with a weak time-varying (often chaotic) external potential.

The reader has already learnt from Dirac perturbation theory that this external "noise" would turn an initially stable ground state of an atom into a superposition of excited (and thus nonstable) quantum states with one or several electron(s) in excited bound state(s). Moreover, such "noise" would even ionize the atom, thus transferring an electron or several electrons from the atomic shell into a classical continuum energy state outside of the range of bound quantum states.

Here we enter into a rich and rapidly developing field of so-called *decoherence* phenomena, which is focused on the problems of the interaction of

quantum systems with their environment. Below we are going to return to
the problem of constraints on lifetimes of gravitational quantum states of
neutrons, atoms and antiatoms, imposed by their interaction with different
types of noise. Some such effects are interesting in themselves, some others
provide useful methodical tools.

The form of integral in the expression (3.23) for a transition amplitude,
although written for small perturbations, inspires a rather natural assumption that the transition amplitude could become large in the case of such
time dependence of the perturbation, $U(z,t)$, that it compensates oscillations of the exponential term, $\exp\left(-i(E_n - E_i)\tau/\hbar\right)$. Such a compen-
sation would occur provided that the perturbation evolves harmonically,
$U(z,t) = u(z)\sin(\omega t)$.

Problem 3.20.   *Verify that the probability to observe a bouncing neutron
in n-th gravitational state $\psi_n(z)$ above a horizontal mirror under the influence of a weak harmonic potential $U(z,t) = u(z)\sin(\omega t)$, provided that
initially the neutron was settled in i-th state $\psi_i(z)$, could be found from the
following expression:*

$$c_n(t) \simeq -iu_{in}\frac{\exp\left(-i(E_n - E_i - \hbar\omega)\tau/\hbar\right) - 1}{E_n - E_i - \hbar\omega} \qquad (3.25)$$

$$-iu_{in}\frac{\exp\left(-i(E_n - E_i + \hbar\omega)\tau/\hbar\right) - 1}{E_n - E_i + \hbar\omega},$$

$$u_{in} = \int_0^\infty \psi_i(z)\psi_n(z)u(z)dz. \qquad (3.26)$$

Our assumption is supported by the fact that one of the denominators
in the above expression could become small provided $E_n - E_i \approx \hbar\omega$ or
$E_i - E_n \approx \hbar\omega$.

In the case of validity of these equalities, the transition amplitude starts
growing linearly in time, as follows: $c_n(t) \approx u_{int}/\hbar$. In the case of the first
equality, $\hbar\omega = E_n - E_i$, the system exhibits transition from a low energy
state i into a high energy state n, i.e. it gets energy from the external field.
In the case of the second equality, $\hbar\omega = E_i - E_n$, the system exhibits the

opposite transition from a high energy state into a low energy state, i.e. it returns energy to the field.

As we already know, the Dirac perturbation theory could be applied only for small transition amplitudes, $c_n \ll 1$; thus we could not use it at too-large evolution times, $t > \hbar/u_{in}$. However, it provides grounds for a reasonable guess that a quantum system would be sucking energy from the field as long as the amplitude of the excited state is smaller than unity; then it would be returning energy to the field and would return to the low energy state, and so on.

In order to verify this hypothesis we invite the reader to derive the result, which is essential for the whole physics of *resonance spectroscopy*, known as the Rabi formula[21].

Problem 3.21. *Assume that a quantum system could be settled only in two quantum states, and it is affected by a weak harmonic potential with the frequency ω: $U(z,t) = u(z)\sin(\omega t)$, with $\Delta\omega = (E_2 - E_1 - \hbar\omega)/\hbar$, and E_1, E_2 being the energies of the quantum states. The system is found initially in the ground state n=1.*

Verify that the equation for the probability to observe this quantum system in the excited quantum state as a function of time $P_{12}(t)$ has the following form:

$$P_{12}(t) = \frac{\Omega}{\Omega^2 + (\Delta\omega)^2} \sin\left(t\sqrt{\Omega^2 + (\Delta\omega)^2}/2\right), \tag{3.27}$$

$$\Omega = u_{12}/\hbar. \tag{3.28}$$

The above expression manifests the fact, which we have already foreseen, that a quantum system performs back and forth transitions between resonantly coupled states.

The frequency of such transitions is equal $\sqrt{\Omega^2 + (\Delta\omega)^2}/2$, and it is known as the *Rabi frequency*. It is important that in the case the frequency of an applied field is very close to the resonance, even a tiny perturbation

---

[21]Rabi, I. I., Millman, S., Kisch, P., and Zacharias, J. R. (1938). The molecular beam resonance method for measuring nuclear magnetic moments. The magnetic moments of Li-3(6), Li-3(7) and F-9(19), *Phys. Rev.* **55**, p. 0526.

turns the system from the ground to the excited state for a sufficiently
long time. In the opposite case, i.e. if the "de-tuning" of the external field
frequency becomes larger than $u_{12}/\hbar$, the maximum transition probability
decreases rapidly.

This result provides us with a very important practical means for measuring the energy difference between states of a quantum system using the
method of induced resonant transition. One should apply an external harmonic field to a quantum system and detect transitions from the initial to
the final state. The necessary condition for exploiting such a method is our
ability to prepare the initial state and to detect the final state.

It's worth mentioning that the typical frequencies of transitions between
the lowest gravitational states of neutrons are of the order of hundreds Hz,
i.e. in the range of the first musical octave. With a certain effort one could
hear the *music of transitions* between gravitational states, or to induce
transitions between gravitational states with music, as shown in figure [3.5].
(Guess the meaning of the snake's shape in this figure.)

Let us see how the Rabi approach of resonant transitions could be used
in the case of a bouncing neutron[22].

First, let us select the lowest gravitational quantum state by means
of transmitting neutrons through a slit between a horizontal mirror and
an absorber, which is positioned at a height $H_1 = \lambda_1 l_0$ above the mirror.
We know already that all states except for the lowest ones are strongly
absorbed and thus they could not pass through the assembly. This is a
so-called preparation stage; it provides that the admixture of "target" final
states into the initially prepared ground state is negligible.

Second, let us apply an external harmonic field with a given frequency
ω. What kind of interaction could we use in order to affect the neutron
motion?

A realistic option is to use the strong interaction, which acts between the
neutron and nuclei in the mirror. Namely, we could make the bottom mirror vibrate with an appropriate frequency. In such a case the neutron would
interact with the optical potential of the mirror, $U(z,t) = U_0\Theta(-a\sin(\omega t))$,
in the required manner. In terms of the analytical description of this problem, one of the boundary conditions would vary: namely the height, at
which the probability to observe neutron is zero, would oscillate.

---

[22]For detailed discussion of neutron gravitational states and their applications see
[Nesvizhevsky, V. V., and Protasov, K. V. (2006). *Quantum states of neutrons in the
Earth's gravitational field: state of the art in Trends in Quantum Gravity Research*,
NOVA Scie. Publ., New York.]

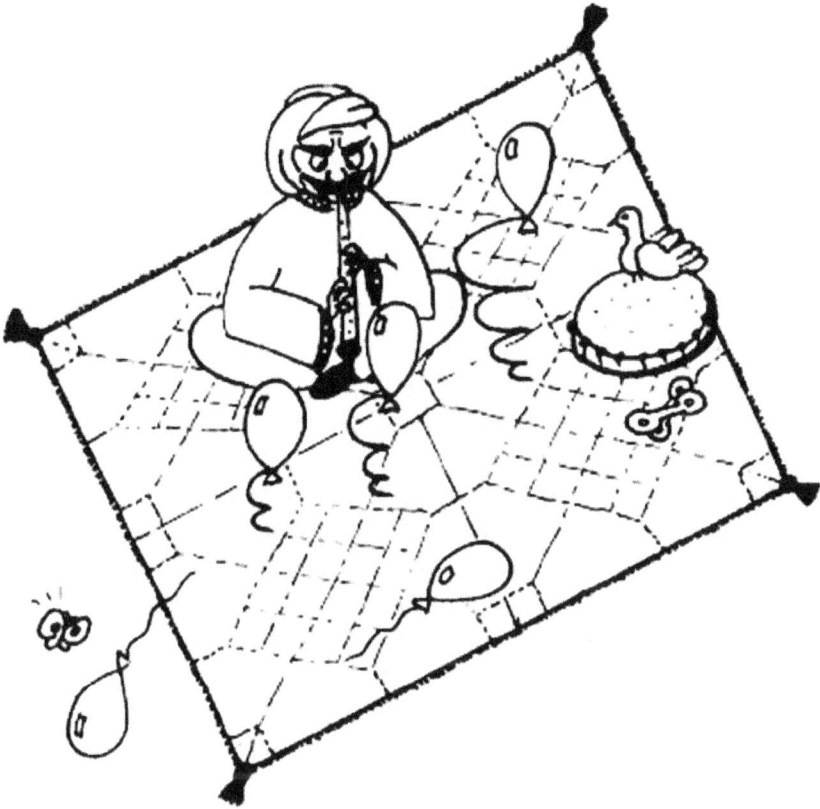

Fig. 3.5   Harmony of transitions.

Another realistic possibility from a longer list of available but less feasible options is to use the interaction of the neutron magnetic moment with a spatially and temporally varying magnetic field, $U(z,t) = -\mu B(z)sin(\omega t)$. A harmonic neutron–field interaction would produce a resonant transition from the initial lowest to the final excited state if only the condition $\omega \simeq (E_f - E_i)/\hbar$ is met, and we also have a means to observe the neutron in the final state f.

Third, we need to have a means to detect the neutron settled in a particular final quantum state in order to establish that the transition has actually happened. This detection could be done in principle using a position-sensitive detector installed at a certain height H_n. Indeed, as we remember, the mesoscopic characteristic spatial size of gravitational quantum state allows us to practically distinguish various states via distinguishing their spatial sizes.

Another related method of detecting the neutron settled in a particular quantum state consists of installing an absorber at the height H_1 above the mirror, so that, for instance, solely the ground gravitational state could pass through the slit. In this case one would measure at the exit of this slit the neutron flux, which contains the information on the probability to depopulate the initial ground quantum state induced by the action of the resonant field of a kind mentioned above.

What are the principal limitations concerning measurements of the resonant transition frequency using this method?

Let us look again, attentively, at the famous Rabi formula (3.28). It is expressed as a product of two terms: the maximum probability of transition and a time-dependent periodic *sin* factor. The first term in the product is interesting for us now; it exhibits a pronounced maximum as a function of the detuning of the frequency from this maximum, $\Delta\omega = 0$. The width of this resonance peak is determined by the intensity of the external perturbation u_{12}.

Thus, the smaller strength of external perturbation, the *narrower* this peak, and thus the more precise estimation of the value of resonance frequency, provided an equivalent level of statistics and the absence of major systematic effects. The second mentioned term in the Rabi expression (3.28) illustrates the fact that a *certain time*, t_m, is required in order to reach the maximum of the transition probability for the first time:

$$t_m = \frac{\pi}{\sqrt{\Omega^2 + (\Delta\omega)^2}}. \qquad (3.29)$$

Thus a universal conclusion, which is valid in particular for all kinds of transitions between gravitational quantum states of neutrons, is the following. If we would like to achieve a better accuracy, then we have to decrease the the intensity of external interaction causing the resonant transition on one hand, but on the other hand we have to increase simultaneously the time of the resonant interaction in order to provide an observable probability of transition.

The reader has probably recognized in the above statement the "hidden" famous *uncertainty relation* for energy and time. In other words, the uncertainty in energy of a neutron in a gravitational quantum state is inversely proportional to the observation time. This is a universal and inevitable statement, which constrains the precision of measurements of energy level spacings in the described type of "flow-through" experiments through a slit.

In such experiments, the time of observation is strictly limited by the time of flight of UCNs through the slit, even if there would be no parasitic or intentional transitions between the states. The typical horizontal velocity of UCNs is a few meters per second, while the typical maximum length of experimental installations is of the order of dozens of cm. Thus, the time t_m could hardly be larger than the 0.1 s that corresponds to the minimum energy width of a resonance line equal to $\Delta E \simeq 10^{-14}$ eV.

Problem 3.22. *How many gravitational quantum states could be resolved using the method of resonance transitions in a "flow-through-type" experiment with a characteristic time of flight of neutrons through the mirror system equal $t = 0.1$ s?*

Note: take into account that two neighboring gravitational states could be resolved provided the resonance line width of each state is smaller than the spacing between these levels.

The formal constraint on the frequency width of a resonance line $\Delta\omega$ does not impose that one could not measure in principle a transition frequency with a much higher accuracy. If such a measurement is repeated several times, then the resulting accuracy would increase accordingly. The actual statistical accuracy of estimating the resonance frequency ω_{if}, which could be achieved in a set of N equivalent measurements, is equal to $\delta\omega_{if} = \Delta\omega_{if}/\sqrt{N}$.

One could improve in this way, at least in principle, the accuracy of measurements of the energy of gravitational level spacing to any desired extent. The reader remembers, however, that any sufficiently large numbers of detected neutrons, N, could be hardly achieved in practice because of relatively low intensity (phase-space density) of UCNs in UCN sources, which we have at hand nowadays or which are discussed in realistic projects for the near future.

However, there is also another potentially dangerous reason for uncertainties in the gravitational resonance spectroscopy. Accounting for *non-resonantly coupled states* shows that an external oscillating field produces not only transitions between states, but also a shift in energy levels. This is a second-order effect, which is proportional to the square of the field intensity. In order to understand its nature we invite the reader to solving the following problem.

Problem 3.23. *Consider a neutron settled in the lowest quantum state in the gravitational field of the Earth. The neutron is also affected by an external harmonic oscillating perturbation with a frequency that is much smaller than the frequency of transition between first and second gravitational states.*

Note: In this case one could apply the adiabatic approach to this problem and assume that the quantum system is settled in the quantum state with parameters, which are following adiabatically the parameters of a slow-varying external field at any moment. Thus one could find the value of the instant energy of the state, which is a function of time.

Calculate such adiabatic (time-dependent) "energy levels" for the lowest gravitational state in the case of a perturbation given in the form $U(z,t) = u_0 z \sin(\omega t)$, provided that the field is weak.

Calculate the second-order correction to the adiabatic energy terms and the mean energy for the period of oscillation $2\pi/\omega$.

We should mention that the energy levels have no strict sense for a system in an external time-dependent field because the energy is not conserved in this case.

However, in a special case, for an external *periodic* time-dependent field, the equation of motion does not change when we perform a time translation by an oscillation period $T = 2\pi/\omega$ of the external field, or by several such periods. In fact, this observation means that the "energy levels" are determined up to $\hbar\omega$; such energy values are known as *quasi-energy* levels. In general terms, they are the energy characteristics of a quantum system in a periodic field.

The practical sense for resonant spectroscopy is in the "position" of a resonance, i.e. in the *frequency of external field* ω_r at which the transition probability gets its maximum value.

Would accounting for the field modification of quasi-energy levels result in shifting the resonant frequency from its "bare" position $\omega_r = (E_f - E_i)/\hbar$?

If so, could such a frequency shift be taken into account? At what precision?

Usually the precision of accounting for such a resonant frequency shift is limited by our knowledge of the intensity and shape of external perturbation, which might be difficult to control with a desired accuracy.

For an enthusiastic reader, we propose to solve an interesting problem concerning a resonance shift caused by a linear periodic perturbation potential.

Study 2. *A neutron is bouncing on a mirror in the Earth's gravitational field. It is affected by an external perturbation field of the form* $U(z,t) = u_0 z \sin(\omega t)$.

Calculate a shift in the resonant frequency of the system caused by this perturbation.

Note: Use the formalism of quasi-energies.

A natural general method of improving the accuracy consists of increasing the *observation time* t_m, or in other terms the time needed to induce the transition. This increase would also allow an experimentalist to decrease the required intensity of the external field, and thus it would allow correspondingly to decrease second-order false effects. It is clear that a significant increase of the time, spent by a neutron in an experimental device of a reasonable size, is hardly possible in the "flow-through" mode.

The way out of this problem is in constructing a *trap*, in which a neutron could bounce in the gravitational states for a sufficiently longer time, being trapped not only in the vertical direction but also in the horizontal direction simultaneously. Such a trap, however, should meet very severe constraints. In particular, any false transitions between gravitational states, which could follow from mixing the vertical and horizontal motions, should be suppressed.

This is a nontrivial problem. Indeed, in order to keep a neutron in such a trap the neutron horizontal velocity should be periodically reversed, for example by means of its reflection from a *vertical wall* of the trap. If the trap vertical wall is inclined even slightly then such a collision would mix a large horizontal velocity component of the neutron and its small vertical component, and thus the intensive parasitic transitions between gravitational states would occur.

Problem 3.24. *A neutron is settled in the ground gravitational quantum state in a specular trap consisting of a horizontal bottom and nearly vertical side walls.*

Calculate the probability of transition of the neutron to an excited gravitational state induced by a sudden collision with the slightly inclined vertical wall.

Estimate the angle of inclination of a vertical wall of the trap, which provides the lifetime of the ground state in the trap equal to 1 s.

However, not only the nonperfect geometrical shape of the trap constrains the maximum achievable lifetime of the neutron in a gravitational quantum state. Another principle problem is the interaction of the neutron with the *environment*. We have already mentioned that the world around us is always full of various oscillating fields and vibrations. Such external interactions are unavoidable. The best one could do is to reduce these perturbations to a certain level.

In experiments with short observation times, the effect of the environment could be neglected at certain conditions. However, with increasing the trapping time, external noise could induce parasitic transitions. The reason for such parasitic effects could be understood as follows: the noise could be thought of as oscillations with different frequencies in a broad range. Among them there could be resonant frequencies, and thus one could expect the enhancement of parasitic effect due to such *resonant contributions*.

Problem 3.25. *A neutron is settled in the gravitational ground state. It is affected by external noise described with the potential $U(z,t) = u_0 z f(t)$, where $f(t)$ is its temporal dependence. The duration of trapping time is 1 s.*

Calculate the probability of the neutron transition to an excited state induced by such noise.

Note: Assume the amplitude u_0 to be small enough in order to provide that the effect of the external noise could be considered as a small perturbation.

Thus it is vital to *shield* the trap from any external temporal perturbations, especially from vibrations of the mirrors as well as from magnetic fields as they are strongly coupled to the neutron.

The project "GRANIT"[23] consists in particular of constructing such a unique specular trap for the long observation of gravitational quantum states. The time of storage of such neutrons, with no parasitic transitions to other states, is planned to be as high as about 1 s at the first stage of this project. This value of the observation time defines the principal limit, which constrains the width of the resonance line; it is a few times 10^{-16} eV.

After this treatise of resonant spectroscopy we conclude that the method of induced transitions, which are based on harmonic oscillating fields with a well-defined frequency, is a promising approach not only for resolving many quantum states but also for measuring precisely the level spacing. In order to achieve the highest possible accuracy, one should meet several important conditions, including a longest possible observation time and a shielding against parasitic perturbations of various origin.

It is also obvious that a resonant perturbation, which we would like to induce on purpose for investigating the gravitational states, should also have a correspondingly narrow line width.

We have tried to convince the reader with this treatise that the neutron is a good example of a quantum bouncing particle. We have even proven that such a system could be considered as a one-dimensional analog of the hydrogen atom, as a relatively simple benchmark system bounded by a gravitational field. It could keep bouncing for a sufficiently long time so that the energies and spatial distributions of such a system could be measured with a relatively high accuracy.

How could we use the information obtained so far?

We are going to discuss some of opportunities in the next sections.

---

[23]Baessler, S., Beau, M., Kreuz, M., Kurlov V., Nesvizhevsky, V. V., Pignol, G., Protasov, K. V., Vezzu, F., and Voronin, A.Yu. (2011). The GRANIT spectrometer, *Comp. Rend. Phys.* **12**, p.707.

Problem 3.26. *For the given inertial mass of a neutron (see, for instance, the Particle Data Group value), how could the value of* gravitational *mass of a neutron be extracted from a measured frequency of transition between two gravitational states?*

Problem 3.27. *Assume that there exists an additional interaction of the type $U(z) = U_0 exp(-z/a)$ between the mirror and the neutron settled in a gravitational quantum state.*

Describe how the frequency of transitions between quantum states would be modified due to the presence of such an interaction.

For what values of the interaction range (a) would the transition frequency be modified in the largest extent (for a given value U_0)?

3.4   Συνοψις

We discussed above the possibility to use *UCNs* as quantum particles bouncing in the gravitational field of the Earth. *Electrical neutrality* of the neutron and its very *small polarizability* are essential features to avoid false effects related to the electromagnetic interactions of neutrons with the surface and environment.

Neutrons are known for their very small scattering cross sections on individual atomic nuclei (about a factor of 10^{-10} smaller than typical interatomic cross sections). This difference explains why neutrons could usually penetrate deep inside the matter bulk. However, the extinction depth is small for UCNs, i.e. for neutrons with velocities of a few meters per second. UCNs are *reflected* from most materials at any incidence angle.

An explanation for this counterintuitive reflection phenomenon is given within quantum mechanics; wave properties of UCNs are essential to understand it. A neutron wave is interacting with many nuclei simultaneously, and thus a coherent contribution is given by an ensemble of atomic nuclei which are located within about the size of the neutron wavelength. For UCNs such a wavelength is much larger than a characteristic interatomic distance. As a result the neutron wave is scattered on the medium in the same way as it would be scattered on a mean effective field with the intensity defined by the atomic density and the individual neutron–nucleon scattering length:

$$U = \frac{2\pi\hbar^2 fn}{m}.$$ (3.30)

Here n is the mean density of nuclei, m the neutron mass, and f the neutron–nuclei scattering length.

For most materials such a potential, known as *optical neutron–nuclei potential*, is repulsive. Its typical value is several tens or a few hundreds neV. Thus UCNs with energies smaller than the optical potential are totally reflected from it. This statement means in particular that one could make a *mirror* for UCNs. Consequently, the neutron could bounce on such a mirror in the gravitational field of the Earth.

Another useful property of the neutron is its relatively small mass, which determines the *mesoscopic* spatial size of the neutron's gravitational states of the order of a few micrometers:

$$l_0 = \sqrt[3]{\frac{\hbar^2}{2mMg}}.$$ (3.31)

This property simplifies the detection of gravitational states.

A method of *scanning* the neutron spatial density was used in the pioneer experiment, which resulted in the discovery of quantum states of neutrons in the gravitational field of the Earth. The idea of this experiment was to pass UCNs through a slit of variable height between a perfect mirror (placed below) and an *absorber* (placed above). A set of gravitational states was formed during the passage of neutrons through the slit.

Neutrons in those states, provided their unperturbed spatial size is larger than the slit size, were penetrating easily into the upper absorber and thus were efficiently lost there. Therefore, only those states with the spatial size smaller than the slit size could pass through the installation unaffected. By means of changing the slit size one could observe *irregularities* in the flux of transmitted neutrons. Such irregularities were revealing as soon as the slit size was approaching the spatial size of another state:

$$H_n \approx l_0(\frac{3}{4}\pi(2n - \frac{1}{2}))^{2/3}.$$ (3.32)

This method allowed measuring the values of classical turning heights H_n of a few lowest gravitational states. It proved the existence of the states on one hand and provided information on their spatial density distribution on the other hand.

However, the principal difficulties await us on the way to significant improvements of the accuracy of evaluating such "quantum steps". The reason consists of the *tunneling* effect, which allows neutrons to penetrate inside the absorber even if the classical turning height is below the absorber height $H_n < H$. It is a pure quantum effect, which is strictly forbidden in classics. It results in *smoothing* the steps in the transition curve, i.e in a smooth dependence of the number of neutrons passed through the mirror/absorber slit as a function of the slit size:

$$\ln F = \exp\left(-4/3(\frac{H - H_1}{l_0})^{3/2}\right) \frac{\tau_{pass}}{\tau_{abs}}. \qquad (3.33)$$

Here F is the transmission factor, H is the slit size, H_1 is the turning height (of the first state), τ_{pass} is the time of passage of the neutron through the installation, and τ_{abs} is the typical absorbtion time, which mainly characterizes the absorber properties. As a result of such washing out, it becomes especially difficult to distinguish neighboring quantum states at large quantum numbers because the states become denser and denser, and the distance between turning heights H_n and H_{n+1} decreases with increasing quantum number n. Practically it is hard to distinguish more than three lowest states.

A way out of this difficulty could consist of producing more effective absorbers. However, this is also a nontrivial task as the absorption efficiency is strongly reduced because of another quantum phenomenon, namely because of *quantum reflection*. The latter phenomenon consists of reflection of a wave from any sharp interface potential. An absolute value of the optical potential of an absorber is usually much larger than the energy of quantum states, thus providing the reflection of neutrons with much smaller energies.

Another drawback of this approach is that *poorly known properties* of the absorber enter into the final expression for the quantum step height, and contribute to the uncertainty of the step height thus limiting the precision of evaluation of any useful information.

Measurements of the energy level spacing between gravitational states is free of the above-mentioned drawbacks. This method could be turned into measuring a very convenient physical property, i.e. an oscillation *frequency*.

A way to turn energy into frequency is to use *resonance* phenomena. It consists of intense *transitions* between initial and final gravitational states if the external harmonic perturbation is applied with a frequency $\omega_0 = (E_f - E_i)/\hbar$, where E_f and E_i are the energies of final and initial states, respectively. The probability of transition from the initial state into the final state is given in Rabi formula:

$$P_{12}(t) = \frac{\Omega}{\Omega^2 + (\Delta\omega)^2} \sin\left(t\sqrt{\Omega^2 + (\Delta\omega)^2}/2\right), \qquad (3.34)$$

$$\Omega = u_{12}/\hbar. \qquad (3.35)$$

Here u_{12} is the intensity of perturbation, and $\Delta\omega = \omega - \omega_0$ is the detuning from the resonance frequency.

The method of resonance transition consists of: i) preparing a pure initial state; ii) inducing a transition from initial to final state; and iii) detecting population of the final state or depopulation of the initial state. As one can see in expression (3.35) the transition probability reveals a sharp maximum at the frequency $\omega = \omega_0$. The smaller the perturbation strength u_{12}, the sharper the resonance. On the other hand, the same expression indicates that time needed to reach the maximum transition probability is equal to $\pi/\sqrt{\Omega^2 + (\Delta\omega)^2}$. Thus in order to get a sharp resonance we have to use *weaker* fields and *longer* observation times. This condition is a manifestation of the time–energy uncertainty relation.

The width of a resonance line is determined by the time that neutron spent in the installation. The best way to improve the accuracy is to make such time longer without significantly losing the number of neutrons. This improvement is possible in specially designed *traps* for UCNs.

An important requirement for such a trap consists of suppressing any kind of perturbation, which could cause *parasitic* transitions between gravitational states. A dangerous source of such transitions could be imperfections of the trap walls. Collisions with such imperfect walls could mix large horizontal and small vertical velocity components between each other. Thus, an inclination of the walls should be controlled to an impressive level to avoid false effects. Another phenomenon, which should be avoided, is vibration of the installation or oscillating magnetic field gradients at certain frequencies. Such vibrational noise could produce parasitic transitions between gravitational states in the case of resonance frequencies in the noise spectrum.

With all the mentioned *precautions*, needed to avoid false effects, one could evaluate a value of the energy spacing between neutron gravitational quantum states. Such a value would be a source of important information about i) the gravitational neutron mass as compared to its inertial mass, and ii) the existence of new fundamental interactions additional to Newton gravity between the mirror and the neutron, with a characteristic range of a few micrometers.

Chapter 4

Bouncing Particles and Their Applications

4.1 Whispering Gallery and Surface Potentials

"No doubt you see the significance of this discovery of mine?"
"It is interesting, chemically, no doubt," I answered, "but practically..."
"Why, man, it is the most practical... discovery for years."

From *A Study in Scarlet* by Arthur Conan Doyle

Now, following Dr. Watson's question addressed to Mr. Sherlock Holmes, it is time to ask ourselves: What is the practical significance of our studies of quantum bouncing particles?

In general terms, the benefit of using a quantum particle instead of a classical bouncing ball is twofold:

i) As far as a quantum particle of interest is tightly bound, it does not absorb or emit the energy of bouncing like a classical ball does. And thus in principle it could keep bouncing in precisely the same quantum state without dissipating or transitioning to other states for a *long time* compared to the characteristic period of one bounce. In contrast to that, the characteristics of motion of a classical bouncy ball are always evolving and after each bounce they are at least slightly different from those before the bounce.

ii) There are means available for *precision diagnostics* of such quantum bouncing, such as the method of resonant transitions or the interference between quantum states, which give us an opportunity to study not only the gravitational mass of the bouncing particle but also any tiny deviations of the potential of interaction of the particle with the surface as well as the potential of interaction of the particle with any other fields, for instance deviating from Newton's law.

Here we are going to discuss in particular the interesting phenomena related to the bouncing of quantum particles, which allow us to study the properties of *surfaces* to a very detailed extent.

We have already mentioned the concept of centrifuge as an important means for understanding phenomena related to inertia and gravity. We have studied a toy example of curved coordinates (that describes space near the surface of a cylinder), in which the effects of gravity were mimicked by this choice of a coordinate system.

Now we are going to investigate an example of a quantum particle, which is bounded near a *curved* material surface with the *centrifugal* force that appears due to motion of the particle along the surface. We are going to show a close relation between this phenomenon and the already obtained results for quantum bouncing of ultracold neutrons (UCNs) in the gravitational field of the Earth, and above that we are going to find out how properties of the surface could be studied using such a quantum particle.

Assume that we have at our disposal a cylindrical tube with the radius R and an ideally polished internal surface. The axis of the cylindrical tube is aligned vertically, i.e. along the direction of the local gravitational field, so that the gravity direction is orthogonal to the cylinder radius. We send a neutron with the velocity v along the internal surface of this tube in such a way that the trajectory of the neutron could be found rather close to the tube surface.

Such a setup could remind you of the famous trick of a motorcyclist riding along a vertical curved wall. The centrifugal force presses the cyclist to the wall. The same centrifugal force keeps the neutron bound to the curved cylindrical surface. Although the axis in the example with a cyclist is horizontal and in our example with a neutron it is vertical, and thus the gravity is directed differently relative to the cylinder axis in these two cases, the nature of force is the same.

Let us turn to the Schrödinger equation for the *radial* neutron motion in such a system:

$$\left[-\frac{\hbar^2}{2m}\frac{\partial^2}{\partial\rho^2} + \frac{\hbar^2\mu^2}{2m\rho^2} + mv_z^2/2 + U(\rho) - E\right]\Psi(\rho) = 0. \qquad (4.1)$$

Here $\Psi(\rho)$ is the neutron wave-function, ρ is the radial distance measured from the cylinder axis, μ is the angular momentum of the neutron motion, $U(\rho) = U_0\theta(\rho - R)$ is the cylinder surface optical potential, which is zero outside the bulk of the cylinder wall and equal to a constant value inside the wall, v_z is the velocity along the cylinder axis, and E is the neutron energy.

Let us make a few comments concerning the above equation.

First, note that due to the cylindrical symmetry of the problem, the *angular momentum* of neutron motion is conserved. Then it is easy to understand the physical sense of the term $\frac{\hbar^2 \mu^2}{2m\rho^2}$ while considering the neutron motion close to the surface so that $\rho \approx R$. Taking into account that the angular momentum $\mu = m v_{||} R / \hbar$, where $v_{||}$ is the component of neutron velocity tangential to the circle of radius R, one could see that for the motion close to the surface $\rho = R$ the centrifugal term is:

$$\frac{\hbar^2 \mu^2}{2m\rho^2} = m v_{||}^2 / 2, \tag{4.2}$$

i.e. it is simply the *kinetic energy* $E_{||}$ of tangential motion.

Second, in the close vicinity of the cylinder surface we could substitute the effective centrifugal potential simply by constant and linear terms as follows:

$$\frac{\hbar^2 \mu^2}{2M\rho^2} \approx m v_{||}^2 / 2 - \frac{m v_{||}^2}{R}(\rho - R). \tag{4.3}$$

The last term in this expression is a *linear potential*, which corresponds to a constant force equal to $m v_{||}^2 / R$. One could recognize in this term the *centrifugal force* acting on the neutron moving in the vicinity of the surface of the cylinder.

It is more convenient to use the deviation from the cylinder surface, $x = R - \rho$, instead of the distance from the cylinder axis in order to describe the neutron position near the cylinder surface. The equation for radial motion in the vicinity of the cylinder surface written in coordinate x takes the following form:

$$\left[-\frac{\hbar^2}{2m} \frac{d^2}{dx^2} + \frac{m v_{||}^2}{R} x + U_0 \theta(-x) - E_r \right] \Psi(x) = 0. \tag{4.4}$$

Here $E_r = E - m v_{||}^2 / 2 - m v_z^2 / 2$ is the radial energy.

The reader could easily recognize that the above equation is similar to the one that describes neutrons bouncing in the gravitational field of the Earth.

An evident difference to that equation is that instead of the gravitational potential Mgx now we deal with the centrifugal interaction $(mv_{\parallel}^2/R)x$. We already had this result in the first chapter while describing the behavior of a ball, which is flying in the vicinity of a curved surface. We showed that the ball motion could be described in curved (polar) coordinates and also that the resulting description is equivalent to the free fall in the gravity field with the strength $\bar{g} = v_{\parallel}^2/R$.

Without any further efforts, the reader could arrive at the conclusion that in the direction perpendicular to the mirror surface there should exist bound quantum states of neutrons in such an effective centrifugal potential, in analogy to the already discussed gravitational quantum states. Thus a *neutron wave* could travel along a curved surface being bound to it with the centrifugal potential from one side and with the optical neutron–nuclei potential of the wall material from another side.

Such a phenomenon is known in acoustics since ancient times. Nowadays it is called the *"whispering gallery"* effect. The phenomenon consists of the localization of a sound wave along curved walls of round galleries, or hemispherical, elliptical, ellipsoidal roofs, etc. A word whispered near such a wall could be clearly heard along the wall at a relatively large distance from the source. St. Paul's Cathedral in London is a place particularly known for the whispering-gallery effect.

This effect there is well pronounced, and that is probably why John William Strutt, better known as Lord Rayleigh, studied there the whispering-gallery phenomenon with a whistle as a sound generator and candles as detectors of sound. Lord Rayleigh established that a sound-wave localization near the wall of the cathedral is the reason for transferring acoustic energy to much larger distances than in the case of free propagation of sound. A curved wall plays the role of a wave-guide.

Lord Rayleigh not only investigated this phenomenon experimentally in St. Paul's Cathedral, but also presented the theory of the whispering-gallery effect in his famous book *The Theory of Sound*[1]. Even if you have never read this book, or it never heard about, you are already acquainted with an essential part of this theory, namely with "gravitational quantum states", i.e. the states of a wave (a quantum particle) in a linear potential.

---

[1]Strutt, J. W. (Baron Rayleigh) (1878). *The Theory of Sound, Vol. 2*, Macmillan, London.

Fig. 4.1 Whispering whales.

Essentially, the same effect explains the long range communication of whales over huge distances near the surface of the ocean, as illustrated in figure [4.1]. Important ingredients of this phenomenon are that the refractive index of salt water increases nearly linearly as a function of depth, and also low frequencies and thus long wavelengths of whale sounds, which are therefore weakly attenuated. With these remarks, the reader could analyze this phenomenon.

Optics is another domain where the whispering-gallery effect is well known and practically used. Thus, optical whispering-gallery resonators are well-polished glass spheres of certain radii. An optical wave could propagate around the inner surface of such a sphere, being localized in the vicinity of the curved wall, just like a neutron wave or a sound wave. An optical wave could wind around the sphere many times if the resonant condition, which depends on the wavelength and the radius of the sphere, is met. Thus, such a sphere plays the role of resonator, which precisely selects a certain wavelength.

Departing from the close analogy between gravitational and centrifugal quantum states of neutrons, let us return to the case of the neutron wave and discuss the physical parameters of a "neutron-wave resonator". One could conclude that the energy scale of whispering-gallery states, i.e.

the states that are bound near a curved surface by a superposition of a centrifugal effective potential and an optical neutron–nuclei potential, is determined by the centrifugal acceleration, $v_{\parallel}^2/R$.

Exercise 4.1. *Verify that the spatial scale and the energy scale of neutron whispering-gallery states could be rewritten as follows:*

$$l_0 = \sqrt[3]{\frac{\hbar^2 R}{2m^2 v_{\parallel}^2}}, \tag{4.5}$$

$$\varepsilon_0 = \sqrt[3]{\frac{\hbar^2 m v_{\parallel}^4}{2R^2}}. \tag{4.6}$$

Thus we have at hand two parameters in order to manipulate the energies of whispering-gallery states: the *tangential* velocity and the cylinder *radius*. This freedom opens an interesting perspective: With a sufficiently large radius of the cylinder, we could use much faster neutrons than UCNs, namely cold neutrons with velocities from about a hundred to a thousand meters per second. This option is interesting as available fluxes of cold neutrons are much larger than those of UCNs, thus we could benefit from better statistics.

The first experiment with the neutron whispering gallery was performed using cold neutrons[2] scattered on a segment of a well-polished silicon cylindrical mirror with the radius of $R = 25.3$ mm. The angular size of the cylinder segment was about 30.5 degrees, so that neutrons in the whispering-gallery mode could be deflected to such large angles, as shown in figure [4.2]. The experiment consisted of observing the angular distribution of neutrons deflected with the mirror.

Let us mention that most of the neutrons from a beam with relatively broad initial angular and spatial distribution passed through the mirror with minimal deflections. This is valid because the value of optical neutron–nuclei potential ($U_0 = 54$neV) was too small compared to the kinetic

---

[2]Nesvizhevsky, V. V., Voronin, A. Yu., Cubitt, R., and Protasov, K. V. (2010). Neutron whispering gallery, *Nature Phys.* **6**, p. 114.

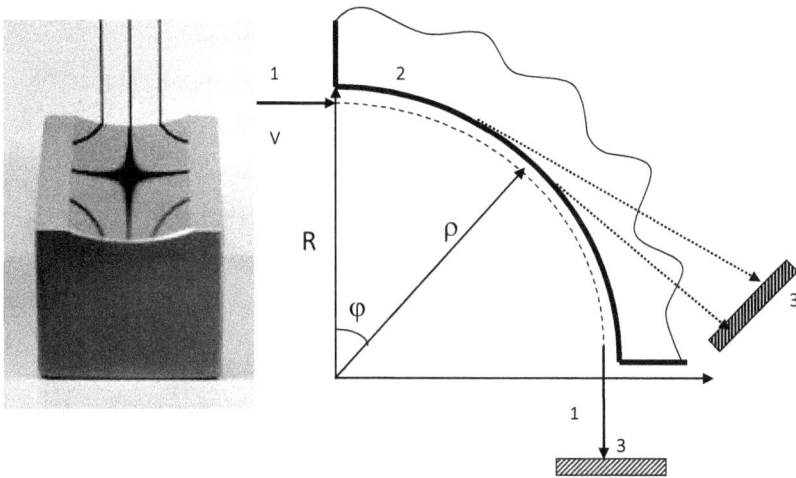

Fig. 4.2   On the left side: A photo of the cylindrical silicon mirror used in actual experiments. On the right side: A sketch of the neutron whispering-gallery experiment on deflecting neutrons with the silicon cylinder mirror. 1 – classical neutron trajectories; 2 – cylindrical mirror; 3 – neutron detector.

energy of a cold neutron. Thus, all neutrons, except for a few captured into whispering-gallery states, pass straight through the bulk of the mirror with minor deflections.

Why does a minor fraction of neutrons captured in whispering-gallery states behave completely differently from all other neutrons? The answer to this question is transparent. In a whispering-gallery state, the kinetic energy corresponding to the neutron *radial* motion is smaller than the value of optical neutron–nuclei potential of the mirror. Note that the major part of the neutron kinetic energy is in its tangential motion, but the tangential motion does not affect the fact of capturing neutrons.

Problem   4.1.   *Estimate the spatial scale and the energy scale of whispering-gallery states of neutrons with the tangential velocity equal to* $v_{||} = 1000$ *m/s for the cylinder radius equal* $R = 25.3$ *mm.*

How many whispering-gallery states could be found in a well, which is produced by a superposition of the centrifugal potential (for the above-given value of the tangential velocity) and the optical neutron–nuclei potential of silicon equal to $U_0 = 54$ *neV?*

Calculate the value of critical velocity, above which no states are possible in the corresponding well.

For given values of the neutron velocity and the mirror radius, what parameter limits the number of whispering-gallery states that could be populated? Obviously, it is the value U_0 of *optical neutron–nuclei potential*. In principle, we have the same problem with gravitational states: neutrons in too-highly excited states could not bounce for a long time on the mirror surface because their energy exceeds the mirror optical potential. However, this happens for extremely excited states in case of gravity.

The situation is different in the case of whispering-gallery states. In the latter case, the centrifugal acceleration could be chosen within a relatively wide range of values. In particular, for certain values of the neutron velocity $v_\parallel$ and the mirror radius R no long-living quantum states become possible at all. For a given neutron velocity this is valid for a sufficiently small mirror radius. For a given mirror radius this is true for a sufficiently large neutron velocity.

By means of decreasing the neutron velocity and/or increasing the mirror radius and thus decreasing the value of the centrifugal acceleration, we could provide conditions that the energy of a whispering-gallery state is approaching to the value of the optical neutron–nuclei potential of the mirror. This is a situation of "appearance of a new quantum state" in the potential well. Close to the barrier, its lifetime is small; however, it increases rapidly with decreasing the centrifugal acceleration.

By means of further decreasing the value of the centrifugal acceleration, we could further decrease the value of the energy of this quantum state below the potential barrier. The *deeper* this state descend, the *longer* the lifetime of the neutrons. Thus we have got flexible and convenient tool for adjusting not only the energy scale and the spatial scale of whispering-gallery states, but also the lifetimes of these states.

Let us investigate this issue in more detail, in order to understand how we could exploit it.

The effective potential well, which bounds neutrons in whispering-gallery states, is shown in figure [4.3].

In this figure, one can observe a triangular potential barrier, which separates the potential well from the left-side area $(\rho > R)$, where the neutron motion is infinite. An infinite motion is a typical condition for the existence of *quasi-stationary* states, i.e. such states that are not "true"

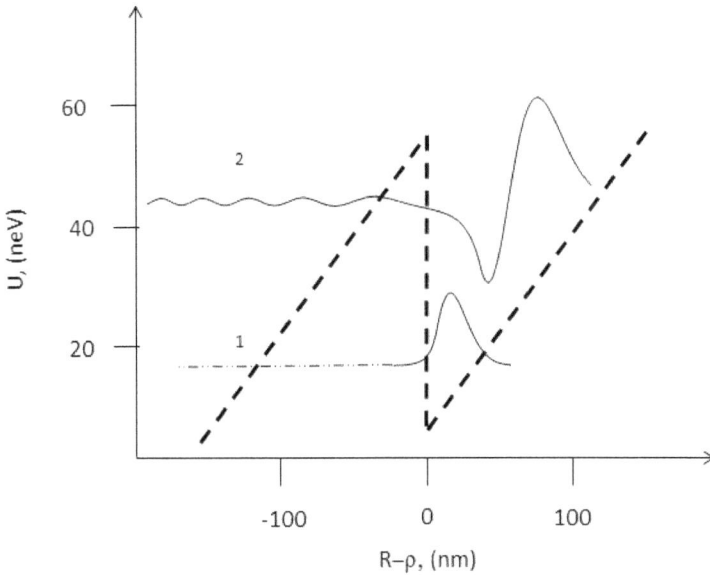

Fig. 4.3 A sketch of the effective potential, which bounds a neutron near the surface of a silicon cylinder, is shown by a dashed line. Wave-functions of quasi-stationary states for ground and first excited states are mainly located within the triangular well; the real parts of these wave-functions are indicated with solid lines. However, these wave-functions are nonzero outside the well and particularly under the barrier, which is due to the decaying character of the quasi-stationary states.

stationary states; they are always decaying during their lifetime[3]. We have already met with quasi-stationary states – these were gravitational states affected by an absorber.

A finite lifetime of such states is associated with the quantum phenomenon of *tunneling*. A neutron could escape from the potential well due to the tunneling through the triangular barrier. A semi-classical expression for the *intrinsic* width of such a state is intuitively transparent, namely it is a product of the frequency of collisions of the neutron with the barrier for a given quantum state and the probability of penetrating through the barrier.

---

[3] For a detailed study of quasi-stationary states and their relation to scattering on curved surfaces see Nussenzveig, H. N. (1992). *Diffraction Effects in Semiclassical Scattering*, Cambridge University Press.

Problem 4.2.   *Calculate the probability of the tunneling of neutrons through the triangular potential barrier.*

Verify that, within the semiclassical approximation, the widths of whispering-gallery states are given in the following expression:

$$\Gamma_n = 2\varepsilon_0 \sqrt{\frac{1}{u_0 - x_n}} \exp\left[-\frac{4}{3}(u_0 - x_n)^{3/2}\right], \qquad (4.7)$$

Here $u_0 = U_0/\varepsilon_0$ is the dimensionless value of optical neutron–nuclei potential of the cylindrical mirror measured in units ε_0, while $x_n \approx (3/2\pi(n - 1/4))^{2/3} - 1/\sqrt{u_0}$ is the semiclassical approximation for the eigenvalues of deeply bound states.

An important message from the above approximate expression is that the width of a whispering-gallery state increases rapidly with increasing the energy of the state, which in its turn increases with the increasing of ε_0. When the energy scale of whispering-gallery states ε_0 approaches the value of the optical neutron–nuclei potential U_0, no long-living states are possible; all states decay faster than the time of the neutrons passage along the mirror length.

As a result, no neutrons would be deflected to large angles. In the opposite case of decreasing ε_0, more and more long-living whispering-gallery states could be formed in the centrifugal potential well and correspondingly more and more neutrons could reach the downstream edge of the cylindrical mirror. Also from this point of view, there is a close analogy to the case of UCNs in gravitational quantum states passing through the slit between an absorber and a flat mirror.

Problem 4.3.   *Derive a precise equation for the energy and the width of neutron whispering-gallery quasi-stationary states.*

Up to this moment we have examined the neutron's radial motion. What about its motion in the tangential and axial directions?

Typical wavelengths, corresponding to these types of motion, are very small compared to the characteristic scales of tangential and axial motion (the cylinder radius, the cylinder length). Thus this type of motion could be treated *classically* with good accuracy. In particular, the tangential motion could be considered as classical motion with a permanent velocity $v_\parallel$. The time of flight of a neutron through a segment of the cylindrical mirror with the angular size φ is thus $t_f = \varphi R / v_\parallel$.

The time of flight could be measured, for instance, by means of interrupting periodically the initial neutron beam in order to shape short neutron pulses. Such short pulses are needed to define the instant of time when the neutrons enter the experimental setup. The *time of flight* measurement is very important in our case because it provides the neutron's tangential velocity, which is a crucial parameter for investigating the whispering-gallery phenomenon.

What kind of density distribution of neutrons deflected by the cylindrical mirror should we expect?

As we have just noticed, the *pattern* should change dramatically as a function of the tangential neutron velocity $v_\parallel$, or in other terms on the neutron wavelength $\lambda = \hbar/(mv)$. Thus, for the neutron wavelengths smaller than a certain critical value λ_c no neutron would be deflected by the cylindrical mirror to the large angles and thus a detector installed at the exit of the mirror, or at the tangential trajectory from this point, would show zero neutron flux.

Problem 4.4. *Consider whispering-gallery states of cold neutrons in the vicinity of a curved cylindrical mirror. The mirror radius is 25.3 mm, the mirror optical neutron–nuclei potential is $U_0 = 54$ neV, the mirror angular size is large enough for shaping the quantum states.*

Estimate the critical neutron wavelength λ_c, below which no neutron could reach a detector placed downstream of the exit edge of the mirror.

Note: Use expression $\Gamma_n = 2\varepsilon_0 \sqrt{\frac{1}{u_0 - x_n}} \exp\left[-\frac{4}{3}(u_0 - x_n)^{3/2}\right]$ (4.7) in order to calculate the lifetime of a whispering-gallery state as a function of the neutron wavelength.

By increasing the neutron wavelength, the lowest whispering-gallery state appears in the triangular potential well (see figure [4.3]). Further continuing in this direction, then the larger the neutron wavelength, the deeper bounding, and thus the larger the lifetime of such a state (compare this dependence to lifetimes of gravitational quantum states as a function of the absorber height). Thus more neutrons could "survive" and reach the detector.

For a certain neutron wavelength, the second quantum state appears in the triangular potential well. Again, with further decreasing of the neutron wavelength, its lifetime increases and neutrons bound in this state could reach the detector. Thus the amount of neutrons that reach the detector increases *irregularly* with the increase of the neutron's wavelength, similar to the case of the transmittance of a variable slit between a mirror and an absorber for neutrons in gravitational quantum states.

However, as soon as two quantum states descend into the potential well, a new interesting phenomenon emerges. Namely, these two states start to *interfere* with each other. We have already discussed the interference of gravitational states in the second chapter of our book, where we established a particularly simple form of interference in the case of two states, given in expression (2.65). The interference of two states manifests itself in a *harmonic evolution* of the probability density of a superposition of these two states. The frequency of "beatings" in the probability density is $\omega_{12} = (E_2 - E_1)/\hbar$.

Problem 4.5. *Build a 3D plot of the neutron probability density for a superposition of two neutron whispering-gallery states as a function of position and time.*

Interference of these states produces a beautiful pattern in a position-sensitive real-time detector placed some distance behind the cylindrical mirror, as is typical for interference phenomena. The positions of strips are defined by properties of whispering-gallery states, including the energy difference between interfering states, their widths and the neutron momentum distribution in these states. Moreover, we could add one more "dimension" to this picture by means of studying it for different neutron wavelengths.

Experimental results obtained in such a way, as well as a simulation of these results using the theoretical model calculations, are given in figure [4.4].

Let us mention that this experimental pattern was measured using the

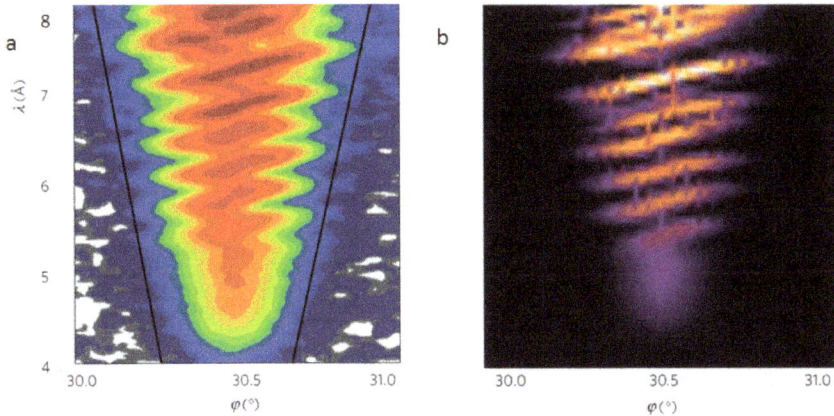

Fig. 4.4   A typical interference pattern at the exit of a cylindrical mirror with the radius of 25.3 mm and the angular size of 30.5° is measured (on the left side) and calculated (on the right side). a – A measured number of neutrons (shown in color) as a function of the deflection angle (in degrees) and the neutron wavelength (in Angstroms). b – A theoretical simulation based on the concept of interference of whispering-gallery states.

time-of-flight method, which allowed us to determine the wavelength of each detected neutron. On one hand, this method provides us with very detailed information concerning the scattering process, which takes place on the mirror surface. On the other hand, this method could be used as a sensitive probe for extra fundamental forces between the neutron and the mirror, as we are going to show in the following section.

Now we are prepared to understand the practical meaning of neutron whispering-gallery states. Any modification of the surface (impurities, an oxide layer, roughness, etc.) would change the potential well, which bounds the whispering-gallery states. Thus the energy difference between quantum levels and the lifetimes would be *modified*, and thus the interference pattern, which we observe, would be modified as well. This fact makes the neutron whispering gallery a very *sensitive tool* for studying any kind of surface effects.

Problem 4.6.   *Build a 3D plot of the neutron momentum distribution density for a superposition of two whispering-gallery states as a function of the neutron radial momentum and the neutron wavelength.*

How would such a distribution be modified, provided the optical neutron–

nuclei potential of a silicon cylinder $U = U_0\Theta(-z)$ is superimposed with an additional potential $U_{ad} = U_1\exp(-z/a)/(1 + exp(-z/a))$ with the following parameters: $a = 10$ nm, $U_1 = 1$ neV?

We have studied the interference pattern that could be measured at a maximum deflection angle, determined by the angular size of the cylindrical mirror. However, there is also another interesting source of information concerning the interference between whispering-gallery states. Neutrons, which *tunnel* through the triangular potential barrier into the bulk of the cylinder, could be detected as well. Let us examine the behavior of such tunneling neutrons in more detail.

First, let us derive the neutron current through the triangular barrier into the bulk of the mirror. A quantum mechanical expression for the current is:

$$j(t,x) = \frac{i\hbar}{2m}\left(\Psi(t,x)^*\frac{d\Psi(t,x)}{dx} - \Psi(t,x)\frac{d\Psi^*(t,x)}{dx}\right). \qquad (4.8)$$

Problem 4.7. *Verify that the above expression for the neutron current $j(t,x)$ meets the continuity equation:*

$$\frac{d\rho(t,x)}{dt} + j(t,x) = 0. \qquad (4.9)$$

Here $\rho(t,x) = |\Psi(t,x)|^2$.

Exercise 4.2. *Verify that the neutron current through the triangular potential for a stationary-bound state equals zero.*

In the case of true bound states, the current is always zero and the probability to observe a neutron in the well is constant. However, the whispering-gallery states are not "truly" bound, but instead they are quasi-stationary states. Therefore neutrons are always *"leaking"* from the potential well due to tunneling, even if there would be no other sources of neutron losses. This tunneling results in a non-zero current.

Problem 4.8. *Verify that the neutron current $j(t,0)$ through the triangle barrier for a superposition of two whispering-gallery states is given in the following expression:*

$$j(t,0) = \frac{\Gamma}{\hbar} \exp\left(-\frac{\Gamma t}{\hbar}\right)(1 - \cos(\omega_{12}t)). \qquad (4.10)$$

Here Γ is the width (for simplicity let us assume that two states have an equal width), $\omega_{12} = (E_2 - E_1)/\hbar$.

There are several important issues in expression (4.10).

First, the neutron current (which could be measured in a correspondingly positioned detector) is proportional to the state *width*, i.e. it is proportional to the probability of penetration through the barrier during unit time. Thus for deeply bound states with very small width such current is negligible, while for the short-living states with the energy close to the top of the barrier, this current is significant for the time interval of the order of their lifetime.

Second, the neutron current decreases with time according to the *decay law* $j(0,t) \sim \exp\left(-\frac{\Gamma t}{\hbar}\right)$ which means that the number of neutrons, which tunnel through the barrier at given time, is proportional to the number of neutrons in the well. Thus we could observe tunneling neutrons only for times of the order of the lifetime of a given state, no longer. However, for the case of many states in the well, their lifetimes are much different. This means that at different times we would observe contributions of different states.

Third, the neutron current demonstrates the effect of "*beatings*", which is determined by the difference of state energies. In particular, in the case of two quantum states the beatings are of harmonic shape. Thus at certain moments of time we are going to observe maxima and minima of the current in the detector.

Finally, as the angular position of a neutron is given in formula $\varphi = v_{||}t/R$, we could establish full correspondence between *time and angle*. This relation is very convenient because the angle could be directly related to the position where a neutron hits a detector placed at a sufficient distance from the cylindrical mirror. Respectively, at different angles measured from the direction of initial neutron trajectory, these maxima and minima are evolving with different phases.

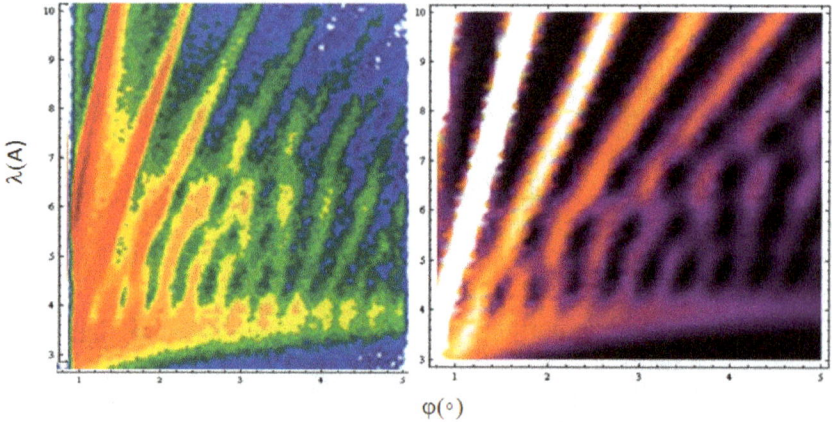

Fig. 4.5 A typical interference pattern for cold neutrons in whispering-gallery quantum states tunneling through the triangular potential barrier. On the left side: A number of neutrons (shown in color) deflected to small angles as a function of the deflection angle (measured in degrees) and the neutron wavelength (in Angstroms). On the right side: The same interference pattern calculated within a theoretical model based on the concept of interference of whispering-gallery states.

In figure [4.5] we present results of measurements of the neutrons tunneling through the triangle potential barrier; such neutrons are deflected to relatively small angles (below 5 degrees). The number of measured neutrons is indicated as a function of the deflection angle and the neutron wavelength. The *"rays"* clearly seen in the figure correspond to the mentioned maxima of neutron deflection to certain angles, with shadows between them corresponding to the minima.

The angles of maximum and minimum deflection depend in their turn, linearly, on the neutron wavelength. This is why we could identify "rays" as the lines of maximum and minimum intensity of deflection. The position and shape of these rays are sensitive to the details of the *neutron–surface interaction*, in particular to the shape of the top of the triangular potential barrier. Indeed, the deflection to small angles is defined by short-living states as we have explained; such states are sensitive to the shape of the *top* of the barrier.

Thus, due to investigations of the distribution of deflected neutrons as a function of the deflection angle and the neutron wavelength, we could distinguish contributions of different whispering-gallery states and observe quantum interference, which manifests itself in pronounced maxima and minima in the count rates of detected neutrons at certain deflection angles. This picture is different for different neutron wavelengths, and it is very sensitive to the properties of neutron–surface interactions.

Within such an approach we could "scan" the properties of neutron–surface interaction in a wide energy range. This option opens an interesting practical perspective for developing precision methods of investigations of surface physics with neutrons, bouncing in whispering-gallery states.

4.2 Fundamental Interactions and Quantum Bouncing

"There is no question that there is an unseen world. The problem is, how far is it from Midtown and how late is it open?"[4]

Woody Allen

Classical physics, based on Newtonian equations, accepted the existence of different kinds of interaction between spatially separated bodies as a matter of phenomenological fact. The main method of classical physics was formulated by Isaac Newton in the anagram "6accdae13eff7i3l9n4o4qrr4s8t12ux"[5], which in a deciphered form and free translation[6] means "given an equation which involves derivatives of one or more functions, find the function".

The idea of classical mechanics is to predict the future (or restore the past) of any simple or complex system by means of decomposing this system into a set of point-like bodies. And then the classical motion of a point-like

---

[4]Allen, W. (1988). *Examining psychic phenomena, in the Ultimate Humour Book,* Chancellor Press, London, UK.

[5]This famous anagram was found in the second letter sent by Isaac Newton to Henry Oldenburg, the Secretary of the Royal Society, where Isaac Newton announced the discovery of calculus on October 24, 1676.

[6]Arnold, V. I. (1984). *Ordinary Differential Equations,* Nauka, Moscow. [English translation Springer-Verlag, Heidelberg, 1987.]

body could be described by means of solving the differential equation of motion provided that the initial conditions and the interactions are given. However, understanding the nature of the interactions themselves was not covered within this concept.

A more ambitious "atomistic" approach aimed at explaining everything in terms of basically free motion of very simple constituents, "atoms" or elementary particles, as we now call them. The first serious step towards understanding the *nature* of interactions was performed in the early twentieth century with the discovery that electrons, atoms and other known light particles exhibit the property of waves. At the same time such "typical waves" like electromagnetic waves exhibit properties of particles, quanta.

This unified wave-particle nature of both particles (electrons) and interaction mediators (photons) assumes that interactions could be described somehow in terms of the motion of elementary particles. This hypothesis turned out to be extremely fruitful when basic ideas of the special theory of relativity were applied to the quantum motion of particles. The greatest success was the development of quantum electrodynamics, which explains electromagnetic interactions and the motion of electrons on the same basis.

Within this concept, the electromagnetic interaction, which had been well established in the frame of description preceding quantum physics, were described in terms of the *"exchange"* of massless particles, photons. Quantum electrodynamics managed to explain not only all those facts, which had been established within classical Maxwell electrodynamics, but also predicted numerous new phenomena, which were confirmed later in experiments.

A natural development of this trend consisted of an attempt to describe all kinds of *fundamental* interactions in terms of an exchange of certain types of elementary particles. The modern Standard Model of elementary particles and fields was developed following this approach. The Standard Model describes strong, weak and electromagnetic interactions in terms of a set of elementary particles (the current total number of elementary particles is 61).

Within such a quantum treatment, an "exchange" is a process, in which a particle mediating a force is created and then absorbed by the interacting-via-exchange particles.

One could show that the typical *spatial scale* of such an exchange interaction is $\lambda \sim \hbar/(mc)$; the scale is determined by the value of mass of the exchange particle. As photons are massless particles, electromagnetic interaction has infinite scale. In contrast, strong interaction is described by the exchange of relatively heavy particles (mesons, with the mass of the order of hundreds MeV), thus they are short-ranged, with a typical radius of a Fermi (10^{-13} cm).

The Standard Model is a very successful theory, its predictions have been verified in many cases including the recent discovery of a very important ingredient of this theory, the Higgs boson. However, there are certain crucial problems, which are not solved within the Standard Model.

The central problem consists of the absence of a unified description for *gravitational* interaction. There is no explanation nowadays for the *accelerating expansion* of the Universe. Another famous problem is the *asymmetry between matter and antimatter*, which we are going to discuss a bit later. There are specific properties of certain particles which have not been fully explained, like *neutrino oscillations*, and some others. The properties of all the particles, the Standard model consists of, are governed by 19 phenomenological parameters, which origin is *not known*.

Among the ideas trying to cover these "gaps" of the Standard Model, there are reasonable hypotheses that predict the existence of *extra light particles* which weakly interact with known matter[7].

What would the existence of such particles change in the world around us? The most evident change consists of an *extra interaction* with a spatial scale defined by the mass of these hypothetical particles. The corresponding scales of such additional interactions could be (according to those theories) in an extremely broad range of distances. Such extra interactions are also predicted in theories with extra spatial dimensions, those explaining dark matter and many others.

Unfortunately we have no crib at our disposal for establishing *a priori* the precise spatial range where one could search for the extra fundamental interactions (figure [4.6]).

Fortunately, we have experimental tools at our disposal, which could appear useful for such a purpose. Thus, all known fundamental interactions enter into the game with a comparable strength for instance for UCNs at typical laboratory conditions. The energy of UCNs in the gravitational field of the Earth is about the same as their energy in a typical laboratory magnetic field, or the optical potential of matter defined by the strong interaction of UCNs with nuclei in the matter. It is also curious to note that the energy scale of gravitational quantum states is about the same as the optical potential splitting due to the weak interaction of UCNs with

---

[7]Beringer, J. *et al.* (Particle Data Group) (2012). *Phys. Rev. D* **86**, p. 010001.

Fig. 4.6   Enigma of forces.

nuclei in the matter of the mirror.

Usually, a discovery of a new elementary particle is associated with huge accelerators and extremely costly mega-projects. However, the fact that such new particles of interest would generate additional (to the already known) interactions could be tested in experiments of a much smaller scale. Perhaps the reader has already guessed that we have in mind quantum bouncing as an option.

Prior to discussing any practical means of using the phenomenon of quantum bouncing to search for new interactions, we have to answer a few questions.

Firstly, if there is indeed a force of a new type, which acts between bodies placed at a certain spatial separation, how could it happen that we have not noticed it yet? The answer could be twofold: either the strength of the interaction of interest is so tiny that the interaction could not be observed within the currently achieved experimental resolution, or this new force has been already effectively included in the interactions; it is a component that has not been distinguished from other interactions.

Thinking on this question might prompt us to propose a route for identifying the phenomena, in which such forces could manifest themselves unambiguously.

An obvious conclusion is that, as long as the precise value of the spatial scale of a new interaction is not known, each characteristic scale of the hypothetical interaction should be probed using a physical system *most sensitive* to the corresponding spatial scale of interaction. Also, we should target at *specific properties* of extra forces, such as their dependence on spin or velocity of the interacting particles. This would allow us to distinguish extra forces from the already known interactions.

Another important point is that most of the existing experimental results were analyzed under certain hypotheses about interaction forces. To verify whether the extra interactions we are looking for have left their traces in the experimental results we have to reanalyze potentially interesting experimental facts under the hypothesis that there is an extra interaction of a new type. This is an important source of information about "new physics", apart from specially designed dedicated experiments.

We know that particles of different types interact in a different manner (electrons participate mainly in electromagnetic and weak interactions, while neutrons and protons participate in strong interactions as well). What particles participating in these hypothetical interactions, we are going to study? The theory predicts different intensity of extra forces depending on the particles involved. Here, it is important for us that neutrons and protons are expected to interact due to the exchange of such hypothetical particles.

In particular, the expected form of interaction of neutron with other nucleons via exchange of such a light hypothetical particle (a group of such light bosons is indicated in the literature by the name "axion" or "axion-like particles"), is as follows:

$$V(\vec{r}) = \hbar^2 g_p g_s \frac{\vec{\sigma} \cdot \vec{n}}{8\pi m_n} \left(\frac{1}{\lambda r} + \frac{1}{r^2} \right) e^{-r/\lambda}. \tag{4.11}$$

Here $g_p g_s$ is the product of constants, which describe the intensity of inter-action (they are called the coupling constants), m_n is the nucleon mass, λ is the range of the force, r is the distance between the neutron and the nucleus, and $\vec{n} = \vec{r}/r$ is the unitary vector. The term $\vec{\sigma} \cdot \vec{n}$ is an indication that the interaction depends on the projection of the neutron magnetic moment on the direction of $\vec{n}$. This is a manifestation of the so-called spin-dependence of extra forces, which could be an important distinguishing feature of such interactions.

Using the known pair neutron–nucleon potential in expression (4.11), one could calculate the potential affecting a neutron in the vicinity of an extended material slab (in the following we will assume that the value of the interaction range λ is much smaller than the characteristic size and thickness of the slab). The reader could perform such a calculation (an integration over all nucleons in the slab) and obtain the neutron-slab potential in the following form:

$$V_a(z) = \pm \frac{g_p g_s}{8} \frac{\hbar^2 \rho_m \lambda}{m_n^2} e^{-z/\lambda}. \tag{4.12}$$

Here ρ_m is the mass density of the mirror material. Signs $\pm$ correspond to different projections of the neutron spin on the direction normal to the surface.

Exercise 4.3. *A neutron is found in the vicinity of a very large material slab; it is affected by an extra interaction acting between the neutron and each nucleus in the slab in the form* $V(\vec{r}) = \hbar^2 g_p g_s \frac{\vec{\sigma} \cdot \vec{n}}{8\pi m_n} \left(\frac{1}{\lambda r} + \frac{1}{r^2} \right) e^{-r/\lambda}$ *(4.11).*

Derive the potential of this extra interaction of the neutron with the slab.

As the force between two bodies is proportional to the product of the number of nucleons in the two bodies involved (and we are evidently inter-ested in increasing the strength of the force to be measured), it is natural to assume that experiments measuring the interaction between macroscopic masses would provide the most sensitive probe for macroscopic interaction ranges, λ, as long as further miniaturization of experimental setups is still feasible.

At even smaller characteristic interaction ranges, at micrometer and nanometer characteristic distances, one of the bodies has to be replaced by an elementary particle or an atom (i.e. by an object with a perfectly defined state), while the second body is still of macroscopic size. A natural condition imposed on the parameters of the probe (an elementary particle) is that its wavelength should be comparable to the characteristic distances to be probed with it.

Such a quantum system with a typical spatial scale from about ten to hundreds of nanometers could be in particular cold neutrons bound in whispering-gallery states, while for a distance from one to several micrometers UCNs in gravitational quantum states could play the role of a "detector" of new interactions.

It is interesting to note that the same tendency in choosing a proper probe for extra interactions continues at even smaller distances. As far as the characteristic distance is comparable to or smaller than a typical inter-atomic distance, one could not discuss macroscopic bodies any longer and the second object has to be replaced by an atom or an elementary particle. But also in this limit, the choice of the optimum particle wavelength is conditioned by the distance to probe, and it is interesting to note that a highly competitive probe is neutron scattering at the nucleus.

Let us note that a "zero" result in an experiment searching for the extra forces in a given distance range with a given sensitivity does not necessarily mean that this experiment has "failed". In contrast to that, if the experiment is more sensitive than all other preceding ones, such a zero result establishes a new limit for the strength of extra forces as a function of the distance range; such additional interactions would be excluded in this area of parameters.

Step-by-step steady exclusion of different scenarios, or different parameter ranges, is often the only possible method of looking for a valid theory. This is particularly true if the parameters are predicted with large uncertainties. Results of searches of this kind are usually presented as *exclusion plots*, i.e. the plot in coordinates "strength of interaction–spatial scale of interaction", which shows the strengths of interaction as a function of the spatial scale, above which extra forces are excluded.

Two examples of exclusion plots relevant to quantum bouncing are pre-

sented in figures [4.7] and [4.8], respectively. These exclusion plots represent
the current state of the art in the field[8].

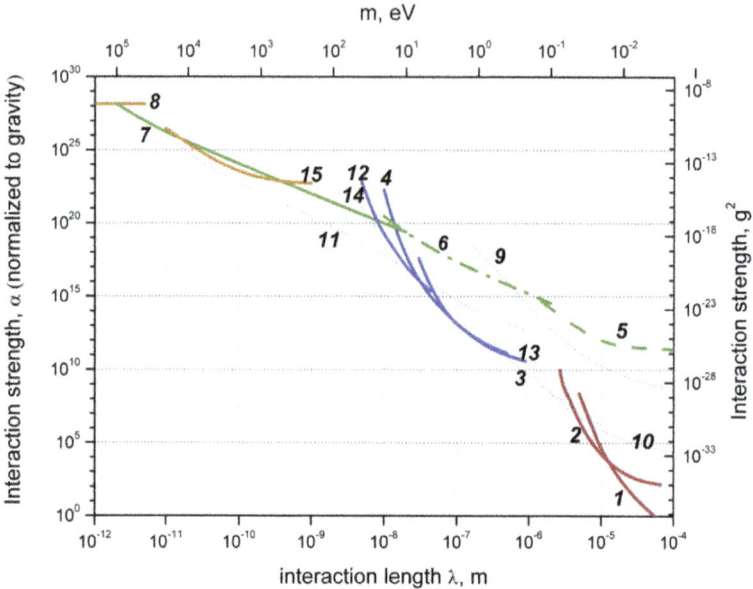

Fig. 4.7   An exclusion plot for spin-independent extra forces. Curves "1" and "2" correspond to "macroscopic" measurements of short-range forces using torsion-balance and cantilever; curves "4","12" and "13" indicate the limits resulting from measurements of Casimir and van der Waals forces between surface and micro-sphere; constraint "7" is derived from analysis of neutron scattering on nuclei; constraint "8" is obtained from spectroscopy of exotic atoms; constraint "15" is obtained from experiments on detection of light bosons from the Sun in germanium detectors; constraints "5" and "6" follow from analysis of first experiments on neutron gravitational and whispering-gallery states, while constraints "9","10" and "11" are estimations of improved limits from new flow-through mode experiments, from GRANIT neutron traps, as well as from whispering-gallery and neutron scattering on nuclei, respectively.

To have an idea of how extra forces would affect quantum bouncing we
propose to solve the following problems.

Let us start with a general question on how the phase of a wave is modified by an additional interaction.

[8]Antoniadis, I., Baessler, S., Buchner, M., Fedorov, V. V., Hoedl, S., Lambrecht, A., Nesvizhevsky, V. V., Pignol, G., Protasov, K. V., Reynaud, S., and Sobolev, Yu. (2011). Short-range fundamental forces, *Compt. Rend. Phys.* **12**, p. 755.

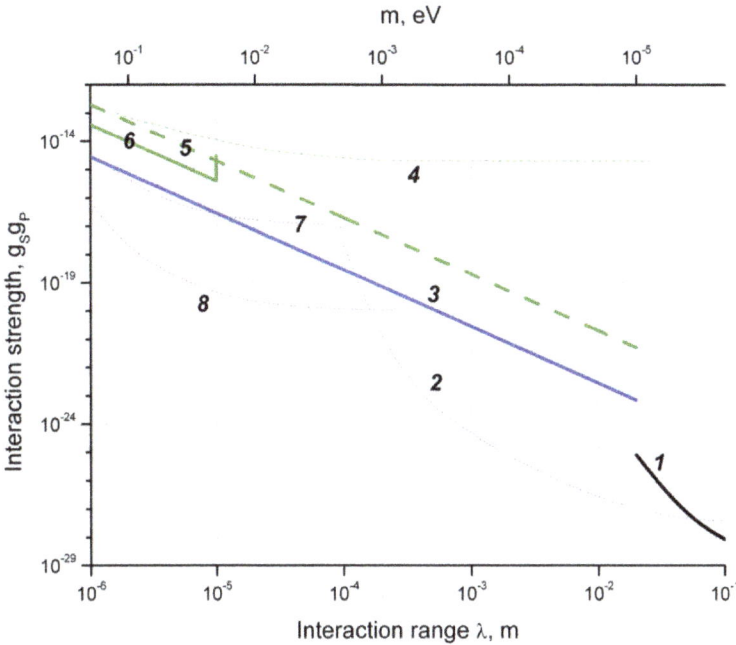

Fig. 4.8   An exclusion plot for spin-dependent extra forces. Limits "2" and "3" are provided by observation of the depolarization rate of polarized He^3; constraint "5" follows from measurements of UCN precession frequency in material traps with different roof and bottom densities; limits "4","7" and "8" are obtained from analyses of UCN gravitational state flow-through experiments, estimation of its improvement with polarized neutrons in flow-through and trap modes in GRANIT, respectively.

Problem 4.9.   *Verify that the phase of a wave, which is propagating with the energy E in a potential $U(x)$, is modified due to the presence of an additional interaction $V(x) \ll U(x)$ according to the following WKB approximation:*

$$\delta\varphi = \int_0^t V(x(\tau))d\tau, \tag{4.13}$$

$$d\tau = dx/v(x). \tag{4.14}$$

Here $v(x) = \sqrt{2(E - U(x))/m}$.

The physical sense of the above result is rather transparent, namely in order to acquire a maximum phase shift a particle should spend in the domain where it is affected by the extra force (within the distance λ from the surface) as long as possible. Thus, results of measurements of such a phase-shift would give us information about extra forces, provided that the benchmark system is well known and any false effects are well controlled.

Such a benchmark system could be a neutron, bouncing in gravitational or whispering-gallery states.

The first attempt to exploit bouncing neutrons for searching for extra forces was a simple analysis of a flow-through experiment in the framework of the hypothesis on the presence of spin-dependent extra forces; such an experiment was discussed in the Chapter 3. Such forces, being attractive for one direction of the neutron spin and repulsive for another direction, should change the transparency of a slit between a mirror and an absorber for a neutron in the function of its spin projection.

This change would result in the "splitting" of a transmission curve (the neutron flux through the slit as a function of its height) into two curves corresponding to the spin "up" and spin "down", respectively. This splitting means in particular that an initially unpolarized neutron beam would become polarized after passing between the mirror and the absorber. It would also change the heights of "steps" in the transition curve, corresponding to the appearance of new quantum states between the mirror and the absorber.

To clarify these points we propose the reader solves the following problem.

Problem 4.10. *How would the spatial size of a gravitational state change provided an extra-potential of a type $V_a(z) = \pm \frac{g_p g_s}{8} \frac{\hbar^2 \rho_m \lambda}{m_n^2} e^{-z/\lambda}$ (4.12) is superimposed with the optical potential of the mirror?*

Although this method provided some sensitivity to eventual modifications of the potential well, which bounds neutrons in quantum states, its accuracy is rather limited. There are two principal factors limiting the accuracy: the time of flight of a neutron through the installation is relatively short, and the influence of the absorber is relatively poorly controlled as the interaction of a neutron with an absorber is quite a complex process.

A much more accurate method could consist of measuring the energy difference between gravitational quantum states settled in a zone above a mirror, which is free of an absorber. As a result, such a measurement provides us with a much higher sensitivity to extra forces. The method of resonant transitions could be a promising approach as far as it gives us an opportunity to measure the transition frequency with an *ultimate precision.*

Problem 4.11. *How would the frequency of transitions between neutron gravitational quantum states or neutron whispering-gallery quantum states change provided an extra potential of the form* $V_a(z) = \pm \frac{g_p g_s}{8} \frac{\hbar^2 \rho_m \lambda}{m_n^2} e^{-z/\lambda}$ *(4.12) were superimposed with the optical potential of the mirror?*

Thus the gravitational states of neutrons could be used as a very clean *benchmark system,* which is sensitive to additional interactions in the range from a few (the lowest states) to several dozens (excited states) of micrometers. The way to increase such sensitivity is to increase the *time of observation* of neutrons in gravitational states. This is a goal of the already mentioned GRANIT experiment, in which the trapping time of gravitational states is expected to reach the order of a few seconds at the first stage.

The expected limits on spin-dependent extra forces from such a long measurement are shown in figure [4.8].

The range of scales from a dozen to several hundreds of nanometers should be studied using neutron quantum bouncing of another type, namely whispering-gallery states.

A very useful property of such states is that their spatial scale could be varied by means of changing the neutron wavelength. This property gives an opportunity to "scan" the mentioned range of distances. Due to the whispering-gallery phenomenon we could observe an *interference pattern,* which is a result of the interference of whispering-gallery states. The change of this pattern due to a possible presence of extra forces could be accurately predicted.

Problem 4.12. *How would change the interference pattern composed of two neutron whispering-gallery states measured at the downstream edge of a cylindrical silicon mirror with the radius of R=25.3 mm and the an-*

gular size of 90 degrees, provided an extra potential of a type $V_a(z) =$ $\pm \frac{g_p g_s}{8} \frac{\hbar^2 \rho_m \lambda}{m_n^2} e^{-z/\lambda}$ *(4.12) is superimposed with the optical potential of the mirror?*

Note: *Consider neutrons with wavelengths in the interval from 4 to 6 Å.*

An interesting feature of the interference pattern of three or more states, which has already been mentioned as the revival phenomenon, could provide another promising tool for sensitive measurements.

Briefly, this feature consists of the appearance of a slow modulation of the interference pattern with a period given by the periods of coherent contribution of three (or more) states. Such coherent contribution results in "super-maxima" and "super-minima" in the overall complicated interference picture, separated by a clearly defined period. For three states such a "super-maxima" period is given in the following expression (we propose the reader verifies it as an exercise):

$$T_s = \frac{2\pi}{\omega_{12} - \omega_{23}} \approx 0.02 \text{ s.} \tag{4.15}$$

Why are revival periods interesting?

In our case, the reason is the following. Generally speaking, minor perturbations of the bounding potential result in minor modifications in the energy level spacing of our quantum system, and thus in minor modifications of the corresponding transition frequencies. However, if the levels of interest are specially chosen, the frequencies of transitions between some states could be close to each other, and thus the revival period would hugely increase or decrease due to a minor perturbation.

Thus even minor perturbations of the bounding potential would result in huge modifications in the corresponding *revival periods*, and therefore the value of a specially chosen revival period might be extremely sensitive to such minor perturbations (so-called "division by zero"). Thus the method of quantum revivals, in particular the method of measuring specially chosen revival periods, becomes an additional *excellent tool* for testing hypothetical extra forces.

Problem 4.13. *Derive an expression for the revival period for a system of three gravitational quantum states.*

Estimate the change of such a period caused by the presence of an extra potential of a type $V_a(z) = \pm \frac{g_p g_s}{8} \frac{\hbar^2 \rho_m \lambda}{m_n^2} e^{-z/\lambda}$ (4.12), which is superimposed with the optical potential of the bottom mirror.

Thus bouncing neutrons could play the role of a benchmark system, a kind of "the hydrogen atom", which provides a reference for studying new forces of nature. Relatively *long observation times, well controlled* false effects, applicability of powerful *interference and resonance spectroscopy* methods, clear basis for the *strict mathematical description* of the system – these are the main benefits of using bouncing neutrons for practical fundamental studies.

In the next section we are going to discuss how quantum bouncing could be used to explore an even more exotic domain: the field of antimatter research.

4.3 Bouncing Antihydrogen and Gravity of Antimatter

"I think that the discovery of antimatter was perhaps the biggest jump of all the big jumps in physics in our century."[9]

Werner Heisenberg

"Isn't antimatter what fuels the U.S.S. Enterprise?"

Dan Brown, *Angels & Demons*

Antimatter is a well-known ingredient of science fiction, with the "practical use" ranging from a powerful explosive or an efficient and compact fuel to a means of absorbing all psychological imperfections of a personality. Science fiction writers, even those who are not well acquainted with physics, nevertheless feel intuitively the mysterious power of the antimatter

---

[9]Heisenberg, W. (1973). from "Development of Concepts in the History of Quantum Theory" in Jagdish Mehra (ed.), The Physicist's Concept of Nature, **1972**, p. 271.

concept. What is the most intriguing property of antimatter that inspires writers?

It is neither a huge energy released from the annihilation of antimatter with ordinary matter, nor its exotic and rare occurrence in the surrounding world. It is the magic symmetry with matter that it provides. Being identical and opposite at the same time, this is what makes antimatter so beautiful!

This feeling of beauty and mystery could become even stronger as soon as one realizes that such an exotic and counterintuitive thing as antimatter had been predicted long before its actual experimental discovery, as an immediate consequence of two very general principles. The reader knows them already. These are the principle of relativity and the wave nature of matter. Isn't it strange that one could get the existence of antimatter out of these two statements?

The idea of antimatter was put forward by Paul Dirac in 1928[10] in his attempts to formulate an equation, which would describe as a simple case the free motion of an electron.

Essential properties of this equation should be the account for the wave nature of the electron and the relativistic principle of relativity. The latter principle, as discussed in Chapter 1, means that all physical equations should be covariant (should keep their form) under transformation from one inertial frame to another. Such transformations were established in special theory of relativity and are known as Lorentz transformations (1.10).

The Schrödinger equation is no longer suitable for a rapidly moving electron as it is not covariant under Lorentz transformations. To find an equation of a proper type for the free motion of a relativistic quantum particle, one should take into account that such an equation should be constructed from energy and momentum operators acting on a free wavefunction, as well as from mass. A Lorentz-invariant combination of these physical quantities is given in the following expression:

$$E^2 - p^2 c^2 = m^2 c^4. \tag{4.16}$$

Here c is the speed of light in vacuum.

This equation results in the famous formula $E_0 = mc^2$ in case the momentum is zero.

A corresponding operator equation could be obtained by means of substituting classical physical quantities by quantum mechanical operators.

---

[10]Dirac, P. A. M. (1928). The Quantum Theory of the Electron, *Proc. Royal Soc. Lond. A* **117**, p. 610.

For the simplest case of a so-called scalar particle, i.e. such a particle that its wave-function does not change under Lorentz transformations, this substitution is already familiar to the reader: $E \to i\hbar d/dt$, and $\vec{p} \to -i\hbar\vec{\nabla}$. The corresponding equation for the free motion of a scalar particle looks like this:

$$\left(-\hbar^2 \frac{d^2}{dt^2} + \hbar^2 c^2 \nabla^2 - m^2 c^4\right) \psi(t, \vec{r}) = 0. \qquad (4.17)$$

Though the case of an electron, which was originally studied by Paul Dirac, is complicated by the fact that electron has an additional degree of freedom, the so-called spin, the structure of the equation of motion is exactly the same.

The above simple-form equation has a plane-wave-type solution:

$$\psi = \exp\left(\pm iEt/\hbar\right) \exp\left(-i\vec{p}\vec{r}/\hbar\right). \qquad (4.18)$$

The problem (in fact, the discovery!) arises from those solutions, which correspond to the "negative energy". They evolve in time like $exp(iEt/\hbar)$. On one hand, the energy of a free particle is always positive. On the other hand, there are solutions of wave equation (4.17), which have both signs of energy. In the case of any interaction with external fields, there is a possibility for transitions between solutions of both types, thus both of them have physical sense.

The guess of Paul Dirac was that these "negative" energy solutions correspond to a new type of particle, counterparts of "ordinary" particles, because they are both the solutions of the same wave equation. They were called antiparticles. It could be shown that the symmetry of the initial equation of the type (4.17) provides that antiparticles have the same mass but opposite charge and magnetic moment to their particle counterparts.

Another conclusion of fundamental importance was that such a relativistic covariant wave equation describes the field, in which the number and type of particles is not conserved. In other words, particles and antiparticles could turn into each other (with restrictions given in conservation laws). These transformations include the famous annihilation – transfer of a pair of particle and antiparticle into other particles (in the case of a slow collision of an electron and a positron they turn into gamma quanta).

The first antiparticle, the positron, was discovered by Carl Anderson[11], a student of the famous Robert Millikan, in a dedicated experiment in 1932, though earlier independent results of Dmitry Skobeltsyn and Chung-Yao Chao (1929) indicated the existence of a particle with a mass equal to the electron mass but with an opposite electric charge. The sources of positrons were cosmic rays, the rays of energetic particles that arrive from interstellar space.

Cosmic rays and thus positrons are always flying around us with a huge velocity, though in relatively small amounts. The traces of positrons even can be found in old photos made by Frédéric and Irène Joliot-Curie. "What counts, however, in science is to be not so much the first as the last" – this comment of famous biochemist Erwin Chargaff is also valid in physics. Anderson discovered positrons without any knowledge of the Paul Dirac predictions.

An experimental proof that the newly discovered particle was indeed the particle predicted by Dirac was made a year later. The experiment was performed by Patrick Blackett and Giuseppe Occhialini[12], who observed the transformation of particles of one type into particles of another type. They discovered the production of pairs of an electron and a positron from gamma radiation that interacted with lead atoms; just as predicted by Paul Dirac.

Since that time the number of newly discovered antiparticles has increased significantly. The list of antiparticles includes, for instance, antiprotons and antineutrons, antineutrinos and a large number of antimesons. Thus the crucial idea that every particle has to have its anti-counterpart, as a consequence of the above-mentioned two basic principles, has been convincingly confirmed.

We are not going into any further detail of the field of physics known as relativistic quantum field theory. It is still under construction. Its most problematic part, which is still to be completed, consists of the need to match it with the general theory of relativity, i.e. in including gravity in the unified theory of interactions on the same basis as other quantized fields are included.

Our aim was just to demonstrate briefly that such a strange and exotic thing as antimatter appears formally in as simple a way as a second root of a square equation. In fact it is a result of implying very general principles of motion to quantum particles.

---

[11]Anderson, C. D. (1932). The positive electron, *Phys. Rev.* **43**, p. 491.
[12]Blackett, P. M. S., and Occhialini, G. P. S. (1933). Some photographs of the tracks of penetrating radiation, *Proc. Royal Soc. London A* **139**, p. 699.

The most strange and exotic thing in antimatter is that it *seems* to us to be strange and exotic. This "seemingness" is due to the fact that the world around us is made of ordinary matter. There have not been y significant amounts of antimatter found in the universe (not taking into account secondary antiparticles, which were born in collisions of energetic cosmic rays). This (non)observation is in direct contradiction to the just established fact of deep symmetry between particles and antiparticles.

This problem of the asymmetry between the amounts of matter and antimatter in the Universe, observed in spite of a symmetry between individual particles and antiparticles, is one of the deep unsolved problems in modern physics. It includes questions concerning the evolution of the Universe, the properties of main particles that take part in weak, strong and electromagnetic interactions known as the Standard Model, and probably the question of gravitational properties of antimatter.

After this short introduction to the mysterious world of antimatter, the reader could ask what, if anything, antimatter has to do with quantum bouncing.

The answer is as simple as gravity.

Indeed, gravitational properties of antimatter are something that one would certainly like to compare in direct experiments with those of matter. Though we have at hand relatively large amounts of charged antiparticles like positrons and antiprotons, their electric charge is a great obstacle on the way to measuring their gravitational properties. We know that even the smallest unavoidable electric fields result in forces much stronger than gravity acting on such particles.

The way out of this problem is to use *neutral antiparticles*. A very good candidate could be a neutral atom, made of an antiproton and a positron, i.e. an *antihydrogen atom*.

We have already learnt that quantum particles should be cold, very very cold, in order to become useful for experimental gravitational studies. Otherwise their gravitational interaction would be too weak to be actually measured. Thus we would like to have *ultracold* antihydrogen atoms with temperatures of at least as small as the typical temperature of UCNs. Cooling antihydrogen atoms to such low temperatures is an extremely nontrivial problem.

While positrons are always present in cosmic rays, and even heavier antiprotons could be generated, for instance, in collisions of highly energetic cosmic protons with media atoms, the probability to produce a neutral antihydrogen atom in collisions of such a kind is rather small (note, however, that this probability is not equal to zero; moreover, first-observed atoms of antihydrogen were produced in the laboratory in collisions of antiprotons with xenon atoms[13]).

In any case there is little chance to catch such an "occasionally" born antihydrogen atom before it is annihilated in collisions with media atoms. Also, such an atom is too hot for our purposes. The way out is to create antihydrogen in the lab by mixing pre-cooled antiprotons and positrons in specially designed traps. During such mixing, the extra energy of positrons could be transferred to surrounding particles and thus positrons could be bound by antiprotons.

An extensive project on synthesis and accumulation of cold antihydrogen atoms is being performed in CERN (The European Organization for Nuclear Research).

This task is extremely challenging; it assumes, in particular, combined coordinated efforts of a large community of best experts in production and cooling of antiprotons and positrons, also in traps of charged particles, detectors, and so on. We are not going into details. We only mention that recent progress in these experimental efforts allows us hope that ultracold antihydrogen atoms are going to become available in large amounts and at phase-space density suitable for making experiments in the near future[14]. Among different projects aimed at studies of antihydrogen we would be specially interested in studies of gravitational properties of falling antiatoms[15,16].

The reader could still be confused.

Indeed, what is in common between an ultracold antihydrogen atom and the phenomenon of quantum bouncing, which we have discussed so many times in this book?

Isn't it the annihilation that instantly destroys an antiatom as soon as it touches any material mirror?

One could easily imagine measurements of the time of fall of an antihydrogen atom in the gravitational field from the moment of its release from an electromagnetic trap to the moment of its annihilation in some material plate underneath. However, one could hardly believe in the reality of

---

[13]Baur, G. *et al.* (1996). *Phys. Lett. B* **368**, p. 251.

[14]Andresen, G. B. *et al.* (ALPHA)(2011). *Nature Phys.* **7**, p. 558.

[15]Perez, P. and Sacquin, Y. (2012). The GBAR experiment: gravitational behavior of antihydrogen at rest. *Class. Quant. Grav.* **29**, p. 184008.

[16]Doser, M. *et al.* (AEgIS Collaboration) (2012). Exploring the WEP with a pulsed cold beam of antihydrogen. *Clas. Quant. Grav.* **29**, p. 184009.

antihydrogen bouncing! "If antimatter and matter make contact, both are destroyed instantly!" as heroes of Dan Brown's books say.

This very natural expectation that an antiatom annihilates immediately after being found in close contact with ordinary matter, surprisingly turns out to be an illusion! This expectation remains valid with a high accuracy for relatively fast antiatoms, for those that we could have at hand in most experiments. However, it conflicts completely with the reality for ultra-cold antiatoms! It is worth investigating this counterintuitive statement in detail.

In fact, the reader (without probably recognizing) has already learnt the answer to this puzzle!

Fig. 4.9   Repulsion of attraction.

It is the phenomenon of *quantum reflection* of a slow atom from a steep *attractive* antiatom-wall interaction potential[17].

---

[17]For a detailed discussion of quantum reflection see Friedrich, H., Jacoby, G., and Meister, C.G. (1997). Quantum reflection by Casimir van der Waals potential tails, *Phys. Rev. A* **65**, p. 032902.

We have discussed this phenomenon in relation to UCNs, which are also reflected from the attractive potential of an absorber rather than penetrating inside it. This effect is purely of quantum nature; it does not exist within classical physics. Surprisingly, in quantum mechanics a deep attractive potential could reflect an incoming wave-particle, i.e. produce a repulsive effect (figure [4.9]).

The following problem could help an interested reader to estimate the main properties of this phenomenon.

Problem 4.14. *Verify that the amplitude of reflection of an antiatom with the energy E and the mass m, which is incident from right to left on a potential well $U(x) = -U_0 \Theta(-x)$, is given in the limit $E \ll U_0$ in the following expression:*

$$S \simeq 1 - 2i \frac{\sqrt{2mE}}{\sqrt{2m(E+U_0)}} = 1 - 2i \frac{\lambda_{well}}{\lambda_{in}}. \qquad (4.19)$$

Here $\lambda_{well} = \frac{\hbar}{\sqrt{2m(E+U_0)}}$ is the effective wavelength of the antiatom inside the potential well, and $\lambda_{in} = \frac{\hbar}{\sqrt{2m(E)}}$ is the wavelength of the incoming antiatom.

The more abrupt the spatial variation of some potential, the smaller the value of the ratio $\frac{\lambda_{well}}{\lambda_{in}}$, and therefore the closer to the unit is the amplitude of the reflected wave. Thus, if a slow enough antiatom falls down to an abrupt and deep potential well, such that the corresponding ratio of the effective wavelengths inside and outside the potential is small, it would be effectively reflected from such a potential. This is the essence of the phenomenon of quantum reflection.

We have already mentioned that in the opposite case of the wavelength of an incoming wave changing slowly and being always small compared to the typical scale of potential, the WKB approximation is valid everywhere. This statement means that a solution of the wave equation has everywhere the form of an incoming wave:

$$\Psi(x) \sim \exp\left(\frac{-i}{\hbar} \int_{-\infty}^{x} \sqrt{2m(E + U(x'))} dx' \right). \qquad (4.20)$$

Thus, the above solution consists of only an incoming wave contribution, i.e. the amplitude of the reflected wave is zero. The necessary condition for quantum reflection is the abruptness of spatial potential change compared to the scale of the antiatom wavelength.

One could compare quantum reflection with another "wave" phenomenon. Imagine you drive a car and keep a cup of coffee on your knees (never use hot coffee while trying to perform such an experiment!). If the road is smooth enough (like a good highway), the coffee would arrive safely inside the cup to the destination point. If the road is not smooth, but instead there is a steep well on it, a wave of your coffee would be "quantum reflected" from the cup onto your knees.

To be more serious, one should not think that quantum reflection is an exotic phenomenon, which could be observed only in state-of-the-art experiments. We meet with quantum reflection every time we look out a window. Each time we look out through a window, especially in dark, we observe our optical reflection in the transparent glass. This effect is due to quantum reflection of a light wave from an abruptly changing optical potential of the glass.

The result (4.19) could be generalized to a broad class of other potentials. The scattering property of potentials, which provide constant asymptotics at $x \to +\infty$ and $x \to -\infty$ in the limit of very small scattering energies, could be described just using one complex constant, known as the scattering length.

$$S \approx 1 - 2ia/\lambda_{in}. \tag{4.21}$$

This expression (4.21) is valid for slow particles, such that their wavelength is much larger than the scattering length: $|a|/\lambda_{in} \ll 1$. The imaginary part of this constant value determines the amplitude of reflected wave. Indeed:

$$|S|^2 = 1 - 4\frac{\operatorname{Im} a}{\lambda_{in}}. \tag{4.22}$$

The above expression establishes the fact that quantum reflection is a universal phenomenon. The smaller the imaginary part of the scattering length, the more prominent the quantum reflection. Most potentials do reflect particles if the particles are slow enough.

Now the question we have to answer could be formulated in mathematically strict terms. Namely, what is the *scattering length* of an antihydrogen atom on a material surface?

An antiatom interacts with a surface at distances much larger than its proper size. This interaction is of pure electromagnetic nature. This fact has been known since early years of atomic theory, and it is related to the names of van der Waals and London. The origin of such an interaction could be understood in terms of the interaction of a dipole moment of an antiatom with an induced dipole in the medium, averaged over the electron state of the atom and the medium.

In the case of a hydrogen atom (the same has to be valid for an antihydrogen atom as the symmetry between matter and antimatter is not expected to be violated for electromagnetic interactions) such a potential is nonzero only in the second order of perturbation treatment (in such a treatment the interaction energy is a small correction to the bounding energy of the electrons in the atom). The typical form of this potential, known as the van der Waals potential, turns out to be:

$$V(x)_{vdW} \approx -C_3/x^3. \qquad (4.23)$$

This potential is attractive.

This form of potential is valid for the distances between an atom and a surface significantly larger than the size of an atom (the Bohr radius is equal to $0.5 \cdot 10^{-8}$ cm).

At even smaller distances the overlapping of atomic electrons and those within the medium takes place, and this overlapping modifies the picture of interaction completely. For an antihydrogen atom at such small distances, an intensive attraction between positrons and electrons, as well as between antiprotons and protons, takes place. This attraction results in fast rearrangement of these particles into attracting pairs, which then annihilate rather promptly.

A more interesting fact is that the expression (4.23) is also not valid for sufficiently large distances. The reason for this deviation is related to the final speed of electromagnetic interactions.

Indeed, the interacting dipoles don't feel each other immediately, but after the time x/c, required for an electromagnetic wave to travel from the atom to the surface. Thus it is more correct to speak about an atom, a surface and the electromagnetic field, which are all interacting with each other. Then the atom–surface potential is just the electron energy in such a system. This retardation effect was described by Hendrik Casimir and Dirk Polder in 1948[18]. They showed that it results in faster decay of potential as a function of the atom–surface distance.

$$V(x)_{CP} \approx -C_4/x^4. \tag{4.24}$$

The following expression matching both asymptotics could be used as an approximate analytic expression for the atom–surface potential:

$$V_{vdW/CP} \approx -\frac{C_4}{x^3(x + C_4/C_3)}. \tag{4.25}$$

The typical distance, where van the der Waals potential turns into the Casimir–Polder potential is $l_c = C_4/C_3$. It is determined by properties of the atom, namely by its typical level spacing.

This Casimir–Polder correction is very important for quantum reflection, as far as it affects the *abruptness* of the resulting potential. And, as we remember, the more abrupt the potential change, the larger the efficiency of quantum reflection.

Let us estimate the distance from the surface at which the quantum reflection takes place. We have already explained that the region, where the reflected wave is "born", is that where the WKB approximation starts to fail due to too-high abruptness of the spatial change of potential. Thus we have to analyze the WKB validity condition at different distances from the surface. Roughly speaking, the WKB approximation works well, where the effective wavelength changes smoothly, i.e.:

$$\frac{d\lambda(x)}{dx} \ll 1, \tag{4.26}$$

$$\lambda(x) = \frac{\hbar}{\sqrt{2m(E - U(x))}}. \tag{4.27}$$

[18]Casimir, H. B., and Polder, D. (1948). The influence of retardation on the London–van der Waals forces, *Phys. Rev.* **73**, p. 360.

Problem 4.15. *Verify that the ranges of distances, where the WKB approximation is valid for small enough energies of antihydrogen atoms in the potential* $V_{vdW/CP} \approx -\frac{C_4}{x^3(x+C_4/C_3)}$ *(4.25) are given in the following conditions:* $x \ll C_4/C_3$ *and* $x \gg (c_4/E)^{1/4}$.

The values of coefficients C_3 and C_4 for a hydrogen atom interacting with a perfectly conducting surface are $C_3 = 0.25$ a.u. and $C_4 = 73.62$ a.u. For an antihydrogen atom these values are the same due to the above-mentioned *symmetry* between particles and antiparticles. This observation means in particular that the distance, where the van der Waals behavior (4.23) is replaced by the Casimir–Polder behavior (4.24), for an antihydrogen atom is around 300 times the Bohr radius.

The above numbers demonstrate a very important fact, namely that quantum reflection (i.e. the generation of reflected wave) for the antihydrogen–surface interaction takes place at large (compared to the Bohr radius) distances from the surface. At smaller distances ($x \ll C_4/C_3$) the atom wavelength in the van der Waals potential is small and changes smoothly, thus no reflected wave could be generated at these short distances. At very large distances ($x \gg (c_4/E)^{1/4}$), where the Casimir–Polder potential is significantly smaller than the incident atom energy, the wavelength is practically constant and again there is no quantum reflection.

Thus, quantum reflection takes place in between these characteristic distances. Only that part of the wave, which is not reflected from this region, could penetrate to the bulk of the surface.

The exact values for the hydrogen (antihydrogen) atom–surface interaction could be calculated to a very high level of precision for the atom–surface distances larger than a few atomic sizes. As soon as we get such a potential we could solve the Schrödinger equation for an atom wave impinging on a surface. Thus we could obtain the scattering amplitude, or, for very low energies, at which the interaction is described with a single parameter, the scattering length.

Now, one could ask what kind of interaction we should introduce into this equation to describe the behavior of antihydrogen inside the bulk of the medium.

To what extent would this very complicated interaction contribute to our reflection amplitude?

Surprisingly, annihilation *simplifies* this problem radically.

Indeed, as we have just established, a significant fraction of the antiatom wave for slow enough antiatoms is reflected back from the region, where quantum reflection takes place; this happens at the distances much larger than the Bohr radius. A small fraction of the wave penetrates through the quantum reflection region, accelerates in the short-range part of the van der Waals potential so that its effective wavelength becomes equal to a fraction of the Bohr radius, and enters into the bulk of the medium.

The bulk of the medium plays the role of a soft absorber, as the typical scale of interatomic interaction is of the order of the Bohr radius. Thus the antihydrogen wavelength changes smoothly and acquires an imaginary part, which corresponds to the typical distance (of the order of the Bohr radius), where the antihydrogen atom annihilates totally. Thus no quantum reflection happens at the interface between surface and vacuum and the fraction of the antiatom wave, which penetrates there, is completely lost.

This conclusion means that any details of the short-range antiatom–surface interactions, as well as the antiatom behavior inside medium could not contribute to the reflection amplitude of antiatoms. It is fully determined by the region, where quantum reflection takes place.

One could compare this effect with an attempt to see the bottom of an ocean by looking downwards from the boat sailing on its surface. One could see, in the best case, what happens a few meters below the water surface, while the deep bottom of the ocean is never seen in the darkness of the water depth. Indeed, a light wave is partially reflected in the layers close to the surface while its other part is completely lost, independently of the properties of the ocean bottom.

In order to clarify this issue we propose the reader solves the following problem.

Problem 4.16. *Consider the reflection of a quantum particle, with the mass equal to the antihydrogen mass, from the following potential, $U(x) = -V\theta(a-x) - W/(1+\exp(x/b))$, which mimics the antiatom–surface interaction.*

Assume that the following relations are valid: $V > 0$, $W = W_r + iW_i$, $W_r > 0$, $W_i > 0$ and also $a \gg b \gg \hbar/\sqrt{2mV}$.

Verify that the amplitude of reflection of the particle from the above po-tential changes weakly as a function of $W_r > 0$ and $W_i > 0$.

Verify that the imaginary part of the scattering length in such a poten-tial, under the discussed conditions, is:

$$\mathrm{Im}\, a = -\frac{\hbar}{\sqrt{2mV}}. \tag{4.28}$$

Now we are completely prepared to calculate the amplitude of reflection of an antihydrogen atom from a material surface. This could be done via solving the Schrödinger equation with a precalculated atom–surface potential $V_{vdW/CP}(x)$:

$$\left(-\frac{\hbar^2}{2m}\frac{d^2}{dx^2} + V_{vdW/CP}(x) - E\right)\psi(x) = 0. \tag{4.29}$$

This equation should be complemented with a boundary condition at the surface, which consists of the requirement that only the incoming wave is present there. This boundary condition is a manifestation of the already mentioned fact that no reflected wave is assumed to be produced in the bulk of the medium.

The result of numerical calculations using precise data on the atom–surface interaction $V_{vdw/CP}$ provides the following value for the *complex scattering length of antihydrogen*:

$$a_{CP} = -0.0027 - i0.029 \ \mu\mathrm{m}. \tag{4.30}$$

Problem 4.17. *Derive an analytical expression for the complex scatter-ing length of antihydrogen if the potential $V_{vdW/CP}(x)$ would be given in the approximate expression (4.25): $V_{vdW/CP} \approx -\frac{C_4}{x^3(x+C_4/C_3)}$.*

Calculate a numerical value and compare it with the result (4.30).

For the wavelengths of incident antihydrogen atoms λ_{in} significantly larger than 29 nm (such wavelength corresponds to the energy of an antihydrogen atom equal to 25 neV; by the way, compare it with a typical UCN energy) the *reflection probability* would be close to unit and it is given in the expression:

$$|S|^2 \approx 1 - 4\frac{|\mathrm{Im}\, a_{cp}|}{\lambda_{in}}. \qquad (4.31)$$

With this study we have demonstrated that slow enough antihydrogen atoms would be efficiently reflected from material walls. This situation is somehow similar to the case of UCNs; their reflection from material walls is also a kind of counterintuitive quantum phenomenon (both if their energy is larger or smaller than the optical potential of the wall material). The similarity between antihydrogen atoms and UCNs is in fact deeper that one would think from the first glance, and we are going to discuss it soon.

Problem 4.18. *Calculate from what height an antihydrogen atom "should be dropped" onto a perfectly conducting surface in order to achieve the reflectivity equal to 98%.*

What heights correspond to the reflection probability equal to 50% and 10%?

Here we arrive at the main statement of this section. It is as simple as follows: An antihydrogen atom can "bounce" on a material surface being bound in long-living quantum gravitational states[19].

How long could antihydrogen bounce in gravitational states?

An intuitive guess comes from the following qualitative arguments. The lifetime of a state (in the semiclassical approximation) is equal to the product of the oscillation frequency and the loss probability.

The first value is given in the semiclassical expression:

$$\omega_n \sim (dE_n/dn)/(2\pi\hbar)$$

---

[19]Voronin, A. Yu., Froelich, P., and Nesvizhevsky, V. V. (2011). Gravitational quantum states of antihydrogen, *Phys. Rev. A* **83**, p. 032903.

while the loss probability, as we have established, is:

$$1 - |S|^2 = 4|\operatorname{Im} a_{CP}|\sqrt{2mE_n}/\hbar.$$

Thus, combining both terms together we get:

$$\frac{\Gamma_n}{2} \approx \frac{1}{\pi\hbar}\frac{dE_n}{dn}|\operatorname{Im} a_{CP}|\sqrt{2mE_n}. \tag{4.32}$$

Taking into account the semiclassical expression for the energies $E_n \approx (3/2\pi(n-1/4)^{2/3}\varepsilon_0$ one could conclude that:

$$\omega_n = \varepsilon_0^{3/2}/2\hbar\sqrt{E_n}. \tag{4.33}$$

We propose the reader to verify that a combination of the above results gives:

$$\frac{\Gamma_n}{2} = \varepsilon_0\frac{|\operatorname{Im} a_{CP}|}{l_0}. \tag{4.34}$$

The surprising message, which is encoded in this expression, is that the width of a state does not change as a function of quantum number n, i.e all gravitational states, for which our approximations are still valid, should have *the same width*! This is a consequence of the particular dependence (4.33) of the bouncing frequency on the energy of the state (note, this dependence is valid not only in quantum mechanics, it is the same in classical physics).

The following expression calculates the *lifetime* of gravitational states of antihydrogen atoms bouncing on a material surface:

$$\tau = \frac{\hbar l_0}{2\varepsilon_0|\operatorname{Im} a_{CP}|} = \frac{\hbar}{2Mg|\operatorname{Im} a_{CP}|}. \tag{4.35}$$

For the perfectly conducting surface this lifetime is equal:

$$\tau = 0.1 \text{ s}$$

Is this lifetime long? It should be compared to the typical gravitational time scale $\tau_0 = \hbar/\varepsilon_0 = 10^{-3}$ s. The lifetime of antihydrogen gravitational states is longer by two orders of magnitude than the typical gravitational time, and this is due to the smallness of the ratio $|\operatorname{Im} a_{CP}|/l_0$. The same ratio provides efficient quantum reflection (4.21). Speaking in classical terms, this means that antihydrogen could bounce many times on the surface before it decays due to annihilation.

The qualitative argument used to derive this result could be rigorously proved by means of precise matching of the Airy functions with the solution of the scattering problem at the distances of the order of $|\operatorname{Im} a_{CP}| \ll x \ll l_0$. We propose the reader performs this useful exercise to get the following general result:

Problem 4.19. *Verify that all gravitational quantum states of antihydrogen atoms acquire the same complex shift in energy due to their interaction with a surface:*

$$\Delta E_n = \varepsilon_0 \frac{a_{CP}}{l_0}. \tag{4.36}$$

The above result is very useful. It proves that as far as all energies are shifted by the same value, such a shift disappears in the *transition frequency*.

After all these efforts have been undertaken to understand the surprising ability of antihydrogen atoms to bounce on a material surface in low gravitational states, we can gather in the harvest of practical results.

First, we note that an antihydrogen atom thought of as a quantum bouncing particle is very similar to a bouncing UCN. Indeed, their masses are nearly equal to each other, they are neutral, and in both cases a plane material surface provides a good mirror for them.

Thus all theoretical results, which have been previously obtained for gravitational quantum states of UCNs, could be applied to antihydrogen atoms with minimum modifications. This conclusion is also valid for the developed experimental methods, including, for instance, "flow through-" and "storage-" type experimental techniques, shaping of quantum states and spectra, resonance transitions and interference techniques, use of position-sensitive detectors, and so on.

A physical quantity we would certainly like to measure is the gravitational mass of antihydrogen. We have established that the method of induced resonant transitions between gravitational states is very promising and would allow us to measure gravitational energy level spacing with a high precision. How could we use this knowledge?

We propose the reader to look into Chapter 2, keep in mind the expression for the gravitational energy scale (2.39) and find out how to extract the gravitational mass out of the resonance frequency ω_{fi}:

$$M_{\bar{H}} = \sqrt{\frac{2m\hbar\omega_{fi}^3}{g^2(\lambda_f - \lambda_i)^3}}. \qquad (4.37)$$

Should we particularly care that the effects related to the antihydrogen–surface interaction could generate false effects concerning the estimated value of the gravitational mass of the antihydrogen atom? We have already mentioned that due to the unique property of gravitational states they acquire the same shift (in the first order of small parameter a_{CP}/l_0), and thus the effects of antiatom–surface interaction are *canceled out* and do not contribute significantly to false effects.

Resonance spectroscopy is a very clean and precise method, which could and should be applied even in the case we have limited statistics (i.e. small amounts of ultracold antihydrogen atoms available).

The precision of this method is essentially defined by the value of the resonance width, which is inversely proportional to the time of life of atoms in the quantum states, or by the time of observation. In our case the precision is principally limited by the lifetime of gravitational states (which is of the order of 0.1 s). Thus any methods of increasing this time is of particular interest. Such an increase would simultaneously reduce significantly all second-order effects causing shifts of resonance lines.

Using different materials we could modify parameters of the Casimir–Polder interaction and thus enhance quantum reflection. Calculations show in particular that silica would provide an increase of twice the lifetime (and correspondingly increasing twice the accuracy, while any second-order effects would be reduced by a factor of four). What is the guiding principle in searching for materials (or meta-materials) that could be promising from the point of view of *enhanced* quantum reflection?

The answer is the following. One should try to make the Casimir–Polder potential more steep, and thus to weaken it. The largest known reflection coefficient is predicted for porous media (by decreasing the density of atoms, we decrease and make steeper the Casimir–Polder interaction) as well as for thin slabs. Though there are certain problems concerning practical use of such materials as plane mirrors, the enhancement of lifetimes of antihydrogen gravitational states by a few times seems in principle feasible.

Gravitational states of antihydrogen atoms are *quasi-stationary* due to annihilation. The states decay at any time, though the decay might be relatively slow. We have already discussed the benefits of quasi-stationary states in relation to the neutron whispering-gallery phenomenon. Quasi-stationary states "inform us" continuously about their existence and evolution; interference of such states could be detected "in fly", i.e. during the evolution of such states.

In the case of whispering-gallery states, information about interference was obtained by detecting tunneling neutrons. In the case of antihydrogen atoms we could detect the atoms leaking through the quantum reflection region into the mirror bulk. Such detection is very efficient, as far as gamma quanta and charged particles, which follow antihydrogen annihilation in the medium are easily detectable. If we have a superposition of only two gravitational states, the annihilation rate would exhibit harmonic beatings (4.10).

Problem 4.20. *Verify that for a superposition of N equally populated gravitational quantum states of antihydrogen atoms the rate $R(t)$ of their annihilation in the mirror has the following form as a function of time:*

$$R(t) = \frac{\Gamma}{\hbar} \exp\left(-\frac{\Gamma}{\hbar}t\right) \left[N + 2 \sum_{i,j,i>j}^{N} (-1)^{i+j} \cos\left(\omega_{ij}t + \varphi_{ij}\right) \right]. \qquad (4.38)$$

Here φ_{ij} is the relative initial phase of states i and j.

Beatings, which we could observe in the timings of the annihilation events during bouncing of antihydrogen atoms on a mirror, are defined by the frequencies of transitions between gravitational levels. The interference pattern of several states in time or in space might become particularly sensitive to modifications of the gravitational interaction of antihydrogen.

Indeed, such a pattern has a rather complicated character defined by a superposition of oscillations with different frequencies (4.38). However, there are pronounced maxima and minima at certain instants in the pattern, which correspond to coherent contributions at such moments of two or more terms. We have already discussed this issue in relation with the so-called revival phenomenon. An observation of a revival period in the interference pattern provides a sensitive method for studying extra forces, acting between a mirror and an antihydrogen atom.

Thus we could conclude that antihydrogen is a simple one-dimensional system, bound by gravity and quantum reflection. The level spacing of such a one-dimensional gravitational atom is defined purely by gravity, while effects of the surface are excluded in the first order of a small parameter a_{CP}/l_0. On one hand, annihilation *"purifies"* the physics of antiatom–mirror interaction due to full cancelation of weakly known close-contact effects. On the other hand, it is a tool of continuous *"nondemolishing"* detecting of the behavior of atoms in the gravitational state. These arguments make the bouncing antihydrogen a very promising laboratory for studying gravitational properties of antimatter.

One could ask, however, what the benefit is of using such an exotic approach? Could one use magnetic or electric fields in order to manipulate antiatoms, in particular to prevent them from getting any contact with matter?

The answer is twofold. First, a material mirror could be defined to a much better precision than magnetic or electric fields. In principal, one could control the quality of mirror surface on the scale of an atom. This is not possible with spatial distributions of the fields. Second, quantum gravitational spectroscopy of antihydrogen atoms seems to be the most precise method of measuring gravitational properties of antimatter at the moment. Finally, let's mention the issue of a fifth force, non-Newtonian gravity and other extra interactions, which could be added to the Standard Model. What are the couplings of hypothetic light bosons, mediators of these new forces, with antimatter? There is no other way to answer this question, but to measure antiatoms bouncing on a surface.

4.4 Συνοψις

It is time to summarize the main ideas of this chapter and also to arrive at final conclusions of our book.

We started with a treatise on a bouncing *elastic ball* and revealed important benefits of *periodic* motion for studying subtle physical effects. In particular we discussed what follows from the *equivalence* of gravitational and inertial masses and how it could be tested using a bouncing ball. We realized that this simple system guides us naturally to general concepts of motion such as the *principle of relativity*, the *principle of equivalence* and the *least action principle*.

We analyzed developments of these general ideas in the limit of very light objects, like elementary particles or atoms. We introduced the *quantum bouncing* of a particle on a material surface in the gravitational field of the Earth as a beautiful and simple system, which allows us to point out important general features of quantum motion.

The phenomenon of *interference*, lying in the heart of wave mechanics, clearly manifests itself in the behavior of a quantum bouncing particle. *Quantum motion* itself is nothing else but interference. It is the interference that helped us to realize the deep sense of the least action principle in quantum mechanics and the relation between quantum and classical motions, which turned out to be a particular case of a variety of motions available for a quantum particle.

We found out that interference of waves, in the case of quantum bouncing in a gravitational field, imposes the existence of characteristic *spatial and temporal scales* associated with a particle. The analytical expressions for these values are composed of the gravitational and inertial masses, the intensity of the gravity field and the Planck constant:

$$l_0 = \sqrt[3]{\frac{\hbar^2}{2mMg}}, \tag{4.39}$$

$$t_0 = \sqrt[3]{\frac{2m\hbar}{M^2g^2}}. \tag{4.40}$$

While l_0 is the typical spatial scale of gravitational bound states, t_0 is the typical period of transitions between such states, or the measure of energy spacing between gravitational levels. A certain relation between these spatial and temporal scales is imposed by the *equivalence principle*:

$$t_0 = \sqrt{2l_0/g}. \tag{4.41}$$

This relation could be regarded as a nonrelativistic quantum-mechanical formulation of the equivalence principle.

Quantum bouncing of a particle in the gravitational field of the Earth had been considered as a purely academic exercise until a breakthrough *experiment* with UCNs proved the existence of gravitational quantum states of neutrons. Due to their electric neutrality on one hand and their perfect reflectivity from material mirrors on the other hand, UCNs turned out to be a perfect choice for the challenging project of observing gravitational quantum states for the first time.

After the discovery of gravitational quantum states, *methods* for their precision studies were developed. They include those based on the *scanning* of their spatial density, *interference* and *resonance spectroscopy*. This development opened plenty of possibilities for practical usage of the quantum bouncing phenomenon. The key "practical" properties of neutron gravitational states are their *mesoscopic* spatial size (tens of microns) and their *well-controlled* interaction with mirrors. These properties make such states *long-living* and easy to detect.

We started discussing practical usage of quantum bouncing with a phenomenon, which is deeply related to gravitational quantum states. Namely, this is a localization of a neutron wave in bound states near a surface of a curved mirror, the so-called *whispering-gallery* effect. These are the states bound by the effective *centrifugal* potential and the *optical* neutron–nuclei potential of the mirror. As far as the centrifugal potential could be accurately represented by its linear term near the mirror surface, the motion of interest is a motion in a linear (locally) potential mv^2/Rx, which differs from the gravitational potential only by replacing the intensity of gravity field g with the centrifugal acceleration v^2/R.

We found that properties of whispering-gallery states are similar to those of gravitationally bound neutron states. The *spatial* and *energy* scales of these states are given in:

$$l_0 = \sqrt[3]{\frac{\hbar^2 R}{2m^2 v_\parallel^2}}, \tag{4.42}$$

$$\varepsilon_0 = \sqrt[3]{\frac{\hbar^2 m v_\parallel^4}{2R^2}}. \tag{4.43}$$

An important difference from gravitational states, however, is the *quasi-stationary* character of such states. Their *lifetime* is defined by the probability of tunneling through the centrifugal barrier:

$$\Gamma_n = 2\varepsilon_0 \sqrt{\frac{1}{u_0 - x_n}} \exp\left[-\frac{4}{3}(u_0 - x_n)^{3/2}\right], \qquad (4.44)$$

Here $u_0 = U_0/\varepsilon_0$ is the value of optical neutron–nuclei potential measured in units ε_0, while $x_n \approx (3/2\pi(n - 1/4))^{2/3} - 1/\sqrt{u_0}$ is semiclassical approximation for the eigenvalues of deeply bound states.

These decaying states are continuously sending us information about their "private life" via neutrons *leaking* through the potential barrier. The most interesting fact about this life is the interference of states, which manifests itself in periodic and temporal *"beatings"* of the probability density. In particular for a superposition of only two quantum states, the *tunneling* neutron current exhibits harmonic oscillations defined by the energy spacing between states:

$$j(t, 0) = \frac{\Gamma}{\hbar} \exp\left(-\frac{\Gamma t}{\hbar}\right)(1 - \cos(\omega_{12}t)). \qquad (4.45)$$

Here Γ is the width (for simplicity we study two states with an approximately equal widths), $\omega_{12} = (E_2 - E_1)/\hbar$.

Such beatings could be observed in particular as maxima and minima in the reflection probability as a function of the deflection angle. Indeed this *angle* could play the role of *time* as far as neutron angular motion along the surface with a given tangential velocity $v_{\parallel}$ could be treated classically $\varphi = v_{\parallel}t/R$, while the radial motion is quantized.

A rich spatial *interference pattern* was measured in a position-sensitive detector placed at the downstream edge of a curved mirror. It is defined by the number of interfering states with lifetimes large enough to reach the far edge of the mirror.

What is the practical significance of this beautiful neutron-optics phenomenon? It is a possibility not only to *predict* precisely an interference pattern but also to deduce from eventual tiny *modifications* of such a picture the existence of any kind of additional (to the optical potential) neutron–surface interactions. Such a whispering-gallery neutron interferometer might be a *precision tool* to study such interactions.

This is particularly important in studying *new forces* of nature, which are expected to be originated from the exchange by hypothetical light particles, which weakly interact with matter. Such particles are required in theories that aim to solve the fundamental problems not covered within the Standard Model. Surprisingly, such a simple system as a bouncing neutron could play its role in searching for new forces, the role that usually belongs to huge accelerator facilities.

We show that gravitational neutron states are particularly sensitive to such extra forces in the spatial range from one micrometer to a few tens of micrometers, while whispering-gallery states are perfect for studying spatial scales from one nanometer to one micrometer.

Developed methods for quantum bouncing diagnostics, which include *resonant spectroscopy* and *interferometry*, are the practical means of searching for new interactions in nature. *Modifications* of the interference pattern, *shifts* of resonance lines would be the signals that could verify the existence of extra forces provided any *false effects* are eliminated or taken into account. Long-living character of neutron quantum states, well-controlled neutron–surface interactions, clear mathematical formulation of the theory describing bouncing neutrons – all these arguments make neutron quantum states a *benchmark system*, an ideal laboratory for studying new interactions.

Perhaps the most intriguing way to use quantum bouncing is to study gravitational properties of *antihydrogen atoms* by means of observing their *bouncing* in the gravitational field of the Earth on a material surface. This assumption contradicts common beliefs that any contact of antimatter with ordinary matter should result in prompt annihilation. However, these beliefs turn to be completely wrong for ultracold antiatoms. The reason for their survival in contact with matter is a phenomenon of *quantum reflection*, i.e. over-barrier reflection of a wave from a steep attractive potential.

We show that the atom–surface long-range van der Waals/Casimir-Polder interaction is *steep* enough to provide almost 100 % efficient reflection of ultracold antihydrogen atoms (with temperatures of about nanokelvin) from material surfaces. This reflection allows for the existence of long-living (with the lifetime of the order of $\tau = 0.1$ s or longer) gravitational states of antihydrogen above material surfaces. Their widths, *the same* for all (at least not too excited) gravitational states is given in the following expression:

$$\frac{\Gamma_n}{2} = \varepsilon_0 \frac{|\operatorname{Im} a_{CP}|}{l_0}. \tag{4.46}$$

Here a_{CP} is a complex constant which determines the quantum reflection from van der Waals/Casimir–Polder atom–surface potential.

On one hand, properties of these states are very *similar* to those of neutron gravitational states. On the other hand, these states are *decaying*, with properties analogous to whispering gallery decaying states of neutrons. In particular one could get "on-line" information concerning interference of such states due to a weak annihilation signal with the intensity modulated with the interference beatings.

A very important property of gravitational states of antihydrogen is that they acquire *the same* small shift due to the interaction with material mirror. Thus this effect of the mirror is *canceled out* in the transition frequencies. This property makes gravitational states of antihydrogen especially interesting for *precision* studying gravitational properties of antimatter.

Both methods, the resonance spectroscopy and interference, are available for studying the gravitational motion of antihydrogen. The value of gravitational mass of antihydrogen (which has been never measured) could be extracted in particular from the resonance transition frequencies:

$$M_{\bar{H}} = \sqrt{\frac{2m\hbar\omega_{fi}^3}{g^2(\lambda_f - \lambda_i)^3}}. \tag{4.47}$$

For hypothetical fields of a certain type, including new forces generated by the exchange of hypothetical light bosons, gravitational states are essentially important. Poorly known antimatter–surface interactions with the characteristic distances in the range from nanometers to hundreds of microns could be studied with bouncing antihydrogen atoms using the already developed methods like those in the case of neutrons.

Thus we have at hand a unique system, a bouncing quantum particle with properties defined by the gravity force alone (or by the centrifugal force in the case of a whispering gallery), which clearly exhibits quantum behavior. Properties of such a system are mathematically transparent and could be accurately predicted. These arguments associate a quantum bouncing particle with a "new hydrogen atom" as long as its role in establishing fundamental physical laws is considered.

The phenomenon of quantum bouncing is universal. Free fall and whispering-gallery states of particles and atoms, antiparticles and antiatoms, electrons above the surface of liquid helium in an attractive electrical potential and rainbow, electric current and quarks, conversations of whales in the ocean and many other phenomena. They seem to be completely different, and at first glance they have nothing in common. However, quantum bouncing in a linear potential is an essential and the only concept needed for understanding all of them. We invite the reader to continue a long list of similar physical examples, which are explained in terms of quantum states in a linear potential.

Moreover, not only general features of quantum bouncing, not only the general formalism describing this phenomenon but also various experimental methods and techniques analyzed and developed throughout the book, are common for its various realizations. We can easily point out many such "openings" to a broader world of other related phenomena and applications, which are not covered in the book. We are going to explore them experimentally and theoretically in the future. However, we would be even more happy to know that the reader is encouraged by our book to do that sooner and better than us.

We arrive at the final point of this book. It was probably not an easy read. We hope that those of you inspired enough to overcome all difficulties, who has managed to break through the counterintuitive concepts and multiple problems, really enjoy physics. We would like to share with these encouraged readers our feeling of the joy of physics.

It consists of displacing the illusion that the world around us is in fact the world that we used to see in the past. Physics teaches us that it is completely, radically different. A flat and small world, which an ancient shepherd used to see, turned out to be a sphere traveling around the Sun or more precisely around the center of mass of the solar system, one sphere among many others in the Universe; rigid bodies turned out to be waves with uncertain velocities and coordinates; empty space and homogeneous time turned out to be a strange curved manifold, which is one whole thing with matter included.

The world around us seems to be full of completely different and unrelated phenomena, but it turns out that many (if not all) of them have a similar explanation, like the propagation of whisper in St. Paul's Cathedral and the bouncing of antihydrogen on a material surface.

The truth of physics is always surprising; this truth does not always make us free, but it does make us happy.

Bibliography

Kepler, J. (1619). *Harmonices Mundi*, Johann Planck, Linz, Austria, p. 189.

Galileo (1638). *Discorsi e Dimostrazioni Matematiche, Intorno a Due Nuoue Scienze*, Lowys Elzevir, Leiden, p. 191.

Viviani, V. (1654). *Racconto Istorico della Vita del Sig.r Galileo Galilei, Accademico Linceo, Nobil Fiorentino, Primo Filosofo e Matematico dell'Altezze Serme di Toscana.*

Fermat, P. (January 1, 1662). Letter to Cureau de la Chambre. For the history of this question see Mahoney, M.S., *The Mathematical Career of Pierre de Fermat, 1601–1665*, 2nd edition, Princeton University Press, 1994, p. 401.

Newton, J. S. (1686). *Philosophiae Naturalis Principia Mathematica, Imprimatur S.Perys, Reg.Soc. Praeses, Londini.*

Lagrange, J. L. (1811). *Mecanique Analytique, Courcier* [Cambridge University Press, 2009].

Coriolis, G. (1835). Sur les équations du mouvement relative des systèmes de corps, *Journ. Ecole Polytechn.* **15**, p. 142.

Strutt, J. W. (Baron Rayleigh) (1878). *The Theory of Sound*, Vol. 2, Macmillan, London.

Lorentz, H. A. (1899). Simplified theory of electrical and optical phenomena in moving systems, *Proc. Royal Netherlands Acad. Arts Sci.* **1**, p. 427.

Einstein, A. (1908). Berichtigungen zu der Arbeit: Über das Relativitätsprinzip und die aus demselben gezogenen Folgerungen, *Jahr. Rad. Elektr.* **5**, p. 98.

Mie, G. (1908). Articles on the optical characteristics of turbid tubes, especially colloidal metal solutions, *Annal. Phys.* **25**, p. 377.

Debye, P. (1909). The heliograph of spheres of any material, *Annal. Phys.* **30**, p. 57.

Lord Rayleigh (1914). Further applications of Bessel's functions of high order to the whispering gallery and allied problems, *Philos. Mag.* **27**, p. 100.

Kaluza, T. (1921). The unity problem of physics, *Berl. Ber.*, p. 966.

de Broglie, L. (1924). Recherches sur la théorie dés quanta, *Thesis, Paris*; (1925) *Annal. Phys.* **3**, p. 22.

Schrödinger, E. (1926). Quantisierung als Eigenwertproblem (Zweite Mitteilung), *Annal. Phys.* **384**, p. 489.

Wentzel, G. (1926). Eine Verallgemeinerung der Quantenbedingungen fur die Zwecke der Wellenmechanik, *Zeit. Phys.* **38**, p. 518.

Kramers, H. A. (1926). Wellenmechanik und halbzahlige Quantisierung, *Zeit. Phys.* **39**, p. 828.

Brillouin, L. (1926). La mechanique ondulatoire de Schrodinger: une methode generale de resolution par approximations successives, *Compt. Rend.* **183**, p. 24.

Klein, O. (1926). The quantum theory and five-dimensional relativity theory, *Zeit. Phys.* **37**, p. 895.

Breit, G. (1928). The propagation of Schroedinger waves in a uniform field of force, *Phys. Rev.* **32**, p. 0273.

Dirac, P. A. M. (1928). The quantum theory of the electron, *Proc. Royal Soc. Lond. A*. **117**, p. 610.

Ambarzumian, V., and Iwanenko, D. (1930). Les electron inobservable et les rayons, *Compt. Rend.* **190**, p. 582.

Chadwick, J. (1932). Possible existence of a neutron, *Nature* **129**, p. 312.

Anderson, C. D. (1932). The positive electron, *Phys. Rev.* **43**, p. 491.

Blackett, P. M. S., and Occhialini, G. P. S. (1933). Some photographs of the tracks of penetrating radiation, *Proc. Royal Soc. London A* **139**, p. 699.

Anderson, H., Fermi, E., and Szilárd, L. (1939). Neutron production and absorption in uranium, *Phys. Rev.* **56**, p. 284.

Rabi, I. I., Millman, S., Kisch, P., and Zacharias, J. R. (1939). The molecular beam resonance method for measuring nuclear magnetic-Moments. The magnetic moments of Li-3(6), Li-3(7) and F-9(19), *Phys. Rev.* **55**, p. 0526.

Fermi, E., and Marshall, L. (1947). Interference phenomena of slow neutrons, *Phys. Rev.* **71**, p. 666.

Feynman, R. P. (1948). Space-time approach to non-relativistic quantum mechanics, *Rev. Mod. Phys.* **20**, p. 367.

Casimir, H. B., and Polder, D. (1948). The influence of retardation on the London–van der Waals forces, *Phys. Rev.* **73**, p. 360.

McReynolds, A. W. (1951). Gravitational acceleration of neutrons, *Phys. Rev.* **83**, p. 172.

Einstein, A., Lorentz, H. A., Minkowski, H., and Weyl, H. (1952). *The Principle of Relativity: A Collection of Original Memoirs on the Special and General Theory of Relativity*, Dover Publications, Mineola, New York, p. 111.

Courant, R., and Hilbert, D. (1953). *Methods of Mathematical Physics 1*, Interscience Publishers, New York.

Gol'dman, I. I., and Krivchenkov, V. D. (1957). *Problems in Quantum Mechanics, Gostekhizdat, Moscow.*

Zeldovich, Y. B. (1959). Storage of cold neutrons, *Sov. Phys. JETP* **9**, p. 1389.

Vladimirsky, V. V. (1960). Magnetic mirrors, channels and traps for cold neutrons, *Sov. Phys. JETP* **12**, p. 740.

Kobzarev, I. Y., and Okun, L. B. (1963). Gravitational interaction of fermions, *Sov. Phys. JETP* **16**, p. 1343.

de Haard (Ed.) (1964). *Selected Problems in Quantum Mechanics, Academic Press, New York.*

Leitner, J., and Okubo, S. (1964). Parity charge conjugation + time reversal in gravitational interaction, *Phys. Rev. B* **136**, p. 1542.

Fermi, E. (1965). *A Course in Neutron Physics, in Collected Papers Vol. 2,* University of Chicago Press, Chicago, USA.

Dabbs, J. W. T., Harvey, J. A., Paya, D., and Horstman, H. (1965). Gravitational acceleration of free neutrons, *Phys. Rev. B* **139**, p. 756.

Landau, L. D., and Lifshitz, E. M. (1965). *Quantum Mechanics, A Course of Theoretical Physics*, Vol. 3, Pergamon Press, UK.

Landau, L. D., and Lifshitz, E. M. (1969). *Mechanics, A Course of Theoretical Physics*, Vol. 1, Pergamon Press, UK.

Baz', A. I., Zel'dovich, Ya. B., and Perelomov, A. M. (1969). *Scattering, Reactions and Decay in Nonrelativistic Quantum Mechanics*, Jerusalem, Israel Program for Scientific Translations.

Lushchikov, V. I., Pokotilo, Y. N., Strelkov, A. V., and Shapiro, F. L. (1969). Observation of ultracold neutrons, *JETP Lett.* **9**, p. 23.

Steyerl, A. (1969). Measurements of total cross sections for very slow neutrons with velocities from 100 m/sec to 5 m/sec, *Phys. Lett. B* **29**, p. 33.

Langhoff, P. W. (1971). Schrodinger particle in a gravitational well, *Am. J. Phys.* **39**, p. 954.

Abramowitz, M., and Stegun, I. A. (1972). *Handbook of Mathematical Functions with Formulas, Graphs, and Mathematical Tables*, Dover Publications, New York.

Berry, M. V., and Mount, K. E. (1972). Semiclassical approximations in wave mechanics, *Rep. Progr. Phys.* **35**, p. 315.

Flugge, S. (1974). *Practical Quantum Mechanics I*, Springer-Verlag, Berlin.

Rauch, H., Treimer, W., and Bonse, U. (1974). Test of a single-crystal neutron interferometer, *Phys. Lett. A* **47**, p. 369.

Cole, M. W. (1974). Electronic surface states of liquid-helium, *Rev. Mod. Phys.* **46**, p. 451.

Gibbs, R. L. (1975). Quantum bouncer, *Am. J. Phys.* **43**, p. 25.

Colella, R., Overhauser, A. W., and Werner S. A. (1975). Observation of gravitationally induced quantum interference, *Phys. Rev. Lett.* **34**, p. 1472.

Koester, L. (1976). Veriication of equivalence of gravitational and inertial mass for neutron, *Phys. Rev. D* **14**, p. 97.

Mezei, F. (1976). Novel polarized neutron devices – supermirror and spin component amplifier, *Comm. Phys.* **1**, p. 81.

Luschikov, V. I. (1977). Ultracold neutrons, *Phys. Today* **30**, p. 42.

Stoika, A. D., Strelkov, A. V., and Hetzelt, M. (1978). Upscattering detected as main reason for anomalous loss of ultra-cold neutrons in neutron storage experiments, *Zeit. Phys. B* **29**, p. 349.

Luschikov, V. I., and Frank, A. I. (1978). Quantum effects occuring when ultracold neutrons are stored on a plane, *JETP Lett.* **28**, p. 1978.

Werner, S. A., Staudenmann, J. L., and Colella, R. (1979). Effect of Earth's rotation of the quantum-mechanical phase of the neutron, *Phys. Rev. Lett.* **42**, p. 1103.

Berry, M. V., and Ballas, N. L. (1979). Non-spreading wave packets, *Am. J. Phys.* **47**, p. 264.

Staudenmann, J. L., Werner, S. A., Colella, R., and Overhauser, A. W. (1980). Gravity and inertia in quantum-mechanics, *Phys. Rev. A* **21**, p. 1419.

Hawking, S. W. (1982). The unpredictability of quantum-gravity, *Com. Math. Phys.* **87**, p. 395.

Sears, V. F. (1982). Fundamental aspects of neutron optics, *Phys. Rep.* **82**, p. 1.

Rubakov, V. A., and Shaposhnikov, M. E. (1983). Do we live inside a domain-wall?, *Phys. Let. B* **125**, p. 136.

Rubakov, V. A., and Shaposhnikov, M. E. (1983). Extra space-time dimensions – towards a solution to the cosmological constant problem, *Phys. Let. B* **125**, p. 139.

Ellis, J., Hagelin, J. S., Nanopoulos, D. V., and Srednicki, M. (1984). Search for violations of quantum-mechanics, *Nucl. Phys. B* **241**, p. 381.

Moody, J. E., and Wilczek, F. (1984). New macroscopic forces, *Phys. Rev. D* **30**, p. 130.

Bonse, U., and Wroblewski, T. (1984). Dynamical diffraction effects in noninertial neutron interferometry, *Phys. Rev. D* **30**, p. 1214.

Atwood, D. K., Horne, M. A., Shull, C. G., and Arthur, J. (1984). Neutron phase-shift in a rotating 2-crystal interferometer, *Phys. Rev. Lett.* **52**, p. 1673.

Arnold, V. I. (1984). *Ordinary differential equations*, Nauka, Moscow.

Visser, M. (1985). An exotic class of Kaluza–Klein models, *Phys. Let. B* **159**, p. 22.

Sakurai, J. J. (1985). *Modern Quantum Mechanics*, Benjamin/Cummings, Menlo Park.

Freed, J. H. (1985). Spin waves in spin-polarized H, *Ann. Phys.* **10**, p. 901.

Steyerl, A., Nagel, H., Schreiber, F. X., Steinhauser, K. A., Gahler, R., Glaser, W., Ageron, P., Astruc, J. M., Drexel, W., and Gervais, R. (1986). A new source of cold and ultracold neutrons, *Phys. Lett. A* **116**, p. 347.

Horne, M. A. (1986). Neutron interferometry in a gravity-field, *Phys. B+C* **137**, p. 260.

Hamilton, W. A., Klein, A. G., Opat, G I., and Timmins, P. A. (1987). Neutron-diffraction by surface acoustic-waves, *Phys. Rev. Lett.* **58**, p. 2770.

Balykin, V. I., Letokhov, V. S., Ovchinnikov, Y B., and Sidorov, A. I. (1988). Quantum state selective mirror reflection of atoms by laser-light, *Phys. Rev. Lett.* **60**, p. 2137.

Werner, S. A., Kaiser, H., Arif, M., and Clothier, R. (1988). Neutron interference induced by gravity – new results and interpretations, *Phys. B C* **151**, p. 22.

Weinberg, S. (1989). Precision tests of quantum-mechanics, *Phys. Rev. Lett.* **62**, p. 485.

Bollinger, J. J., Heinzen, D. J., Itano, W. M., Gilbert, S. L., and Wineland, D. J. (1989). Test of the linearity of quantum-mechanics by RF spectroscopy of the Be-9+ ground-state, *Phys. Rev. Lett.* **63**, p. 1031.

Schaerpf, O. (1989). Comparison of theoretical and experimental behaviour of supermirrors and discussion of limitations, *Phys. B* **156**, p. 631.

Antoniadis, I. (1990). A possible new dimension at a few TeV, *Phys. Let. B* **246**, p. 377.

Kasevich, M. A., Weiss, D. S., and Chu, S. (1990). Normal-incidence reflection of slow atoms from an optical evanescent wave, *Opt. Lett.* **15**, p. 607.

Ignatovich, V. K. (1990). *The Physics of Ultracold Neutrons*, Clarendon Press, Oxford.

Kolb, E. W., and Turner, M. (1990). *The Early Universe*, Addison-Wesley, Redwood, CA.

Golub, R., Richardson, D. J., and Lamoreaux, S. K. (1991). *Ultra-cold Neutrons*, A. Higler, Bristol, UK.

Bertolami, O. (1991). Nonlinear corrections to quantum-mechanics from quantum-gravity, *Phys. Lett. A* **154**, p. 225.

Nussenzveig, H. N. (1992). *Diffraction effects in Semiclassical scattering*, Cambridge University Press, Cambridge, UK.

Wallis, H., Dalibard, J., and Cohen-Tannoudji, C. (1992). Trapping atoms in a gravitational cavity, *Appl. Phys. B* **54**, p. 407.

Leeb, H., and Schmiedmayer, J. (1992). Constraint on hypothetical light interacting bosons from low-energy neutron experiments, *Phys. Rev. Lett.* **68**, p. 1472.

Pendlebury, J. M. (1993). Fundamental physics with ultracold neutrons, *Ann. Rev. Nucl. Part. Sci.* **43**, p. 687.

Yu, A. I., Doyle, J. M., Sandberg, J. C., Cesar, C. L., Kleppner, D., and Greytak, T. J. (1993). Evidence for universal quantum reflection of hydrogen from liquid-He-4 , *Phys. Rev. Lett.* **71**, p. 1589.

Aminoff, C. G., Steane, A. M., Bouyer, P., Desbiolles, P., Dalibard, J., and Cohen-Tannoudji, C. (1993). Cesium atoms bouncing in a stable gravitational cavity, *Phys. Rev. Lett.* **71**, p. 3083.

Su, Y., Heckel, B. R., Adelberger, E. G., Gundlach, G. H., Harris, M., Smith, G. L., and Swanson, H. E. (1994). New test of the universality of free-fall, *Phys. Rev. D* **50**, p. 3614.

Mabuchi, H., and Kimble, H. J. (1994). Atom galleries for whispering atoms – binding atoms in stable orbits around an optical-resonator, *Opt. Let.* **19**, p. 749.

Feynman, R. P., Morinigo, F. B., and Wagner, W. G. (1995). *Feynman Lectures on Gravitation*, Ed. Brian Hatfield, Addison-Wesley, USA, p. 11.

Steane, A., Szriftgiser, P., Desbiolles, P., and Dalibard, J. (1995). Phase modulation of atomic de Broglie waves, *Phys. Rev. Lett.* **74**, p. 4972.

Roach, T. M., Abele, H., Boshier, M G., Grossman, H. L., Zetie, K. P., and Hinds, E. A. (1995). Realization of a magnetic-mirror for cold atoms, *Phys. Rev. Lett.* **75**, p. 629.

Baur, G. *et al.* (1996). *Phys. Lett. B* **368**, p. 251.

Lykken, J. D. (1996). Weak scale superstrings, *Phys. Rev. D* **54**, p. R3693.

Onofrio, R., and Viola, L. (1996). Quantum dumping of position due to energy measurements, *Phys. Rev. A* **53**, p. 3773.

Sharratt, M. (1996). *Galileo: Decisive Innovator*, Cambridge University Press, Cambridge, UK, p. 75.

Felber, J., Gahler, R., Rausch, C., and Golub, R. (1996). Matter waves at a vibrating surface: Transition from quantum-mechanical to classical, *Phys. Rev. A* **53**, p. 319.

Bernet, S., Oberthaler, M. K., Abfalterer, R., Schmiedmayer, J., and Zeilinger, A. (1996). Coherent frequency shift of atomic matter waves, *Phys. Rev. Lett.* **77**, p. 5160.

Friedrich, H., Jacoby, G., and Meister, C. G. (1997). Quantum reflection by Casimirvan der Waals potential tails, *Phys. Rev. A* **65**, p. 032902.

Viola, L., and Onofrio, R. (1997). Testing the equivalence principle through freely falling quantum objects, *Phys. Rev. D* **55**, p. 455.

Littrell, K. C., Allmann, B. E., and Werner, S. A. (1997). Two-wavelength-difference measurement of gravitationally induced quantum interference phases, *Phys. Rev. A* **56**, p. 1767.

Vernooy, D. W., and Kimble, H. J. (1997). Quantum structure and dynamics for atom galleries, *Phys. Rev. A* **55**, p. 1239.

Arkani-Hamed, N., Dimopoulos, S., and Dvali, G. (1998). The hierarchy problem and new dimensions at a millimeter, *Phys. Let. B* **429**, p. 263.

Antoniadis, I., Arkani-Hamed, N., Dimopoulos, S., and Dvali, G. (1998). New dimensions as a millimeter to a fermi and superstrings as a TeV, *Phys. Let. B* **436**, p. 257.

Arkani-Hamed, N., Dimopoulos, S., and Dvali, G. (1999). Phenomenology, astrophysics, and cosmology of theories with submillimeter dimensions and TeV scale quantum gravity, *Phys. Rev. D* **59**, p. 086004.

Gea-Banacloche, J. (1999). A quantum bouncing ball, *Am. J. Phys.* **67**, p. 776.

Peters, A., Chung, K. Y., and Chu, S. (1999). Measurement of gravitational acceleration by dropping atoms, *Nature* **400**, p. 849.

Fischbach, E., and Talmadge, C. L. (1999). *The Search for Non-Newtonian Gravity*, Springer, New York.

Nesvizhevsky, V. V., Strelkov, A. V., Geltenbort, P., and Iaydjiev, P. S. (1999). Observation of a new mechanism of ultracold-neutron losses in traps, *Phys. At. Nucl.* **62**, p. 776.

Nesvizhevsky, V. V., Strelkov, A. V., Geltenbort, P., and Iaydjiev, P. S. (1999). Investigation of storage of ultra-cold neutrons in traps, *Europ. Phys. J. Appl. Phys.* **6**, p. 151.

Barabanov, A. L., and Belyaev, S. T. (1999). Potential of ultracold-neutron interaction with matter, *Phys. At. Nucl.* **62**, p. 769.

Nesvizhevsky, V. V., Boerner, H. G., Gagarski, A. M., Petrov, G. A., Petukhov, A. K., Abele, H., Baessler, S., Stoeferle, Th., and Soloviev, S. M. (2000). Search for quantum states of the neutron, *Nucl. Instr. Meth. A* **440**, p. 754.

Rauch, H., and Werner, S. A. (2000). *Neutron Interferometry, Lessons in Experimental Quantum Mechanics*, Oxford Science Publishers, Clarendon, Oxford.

Strelkov, A. V., Nesvizhevsky, V. V., Geltenbort, P., Kartashov, D. G., Kharitonov, A. G., Lychagin, E. V., Muzychka, A. Yu., Pendlebury, J. M., Schreckenbach, K., Shvetsov, V. N., Serebrov, A. P., Taldaev, R. R., and Iaydjiev, P. (2000). Identification of a new escape channel for UCN from traps, *Nucl. Instr. Meth. A* **440**, p. 695.

Mashhoon, B. (2000). Gravitational couplings of intrinsic spin, *Clas. Quant. Grav* **17**, p. 2399.

Barabanov, A. L., and Belyaev, S. T. (2000). Multiple scattering theory for slow neutrons (from thermal to ultracold), *Europ. Phys. J. B* **15**, p. 59.

Belyaev, S. T., and Barabanov, A. L. (2000). Interaction of ultra-cold neutrons with condenced matter, *Nucl. Instr. Meth. A* **440**, p. 704.

Butterworth, J., Geltenbort, P., Korobkina, E., Nesvizhevsky, V. V., Schreckenbach, K., and Zimmer, O. (2000). Proceedings of the International Workshop on Particle Physics with Slow Neutrons, Institut Laue-Langevin, Grenoble, France, October 22–24, 1998. Preface, *Nucl. Instr. Meth. A* **440**, p. V.

Frank, A. I., and Gahler, R. (2000). Time focusing of neutrons, *Phys. At. Nucl* **63**, p. 545.

Obukhov, Yu. N. (2001). Spin, gravity and inertia, *Phys. Rev. Lett.* **86**, p. 192.

Shimizu, F. (2001). Specular reflection of very slow metastable neon atoms from a solid surface, *Phys. Rev. Lett.* **86**, p. 987.

Kohl, M., Hansch, T. W., and Esslinger, T. (2001). Measuring the temporal coherence of an atom laser beam, *Phys. Rev. Lett.* **87**, p. 160404.

Nesvizhevsky, V. V., Boerner, H. G., Petukhov, A. K., Abele, H., Baessler, S., Ruess, F. J., Stoeferle, Th., Westphal, A., Gagarski, A. M., Petrov, G. A., and

Strelkov, A. V. (2002). Quantum states of neutrons in the Earth's gravitational field, *Nature* **415**, p. 297.

Oraevsky, A. N. (2002). Whispering-gallery waves, *Quant. Electron.* **32**, p. 377.

Nesvizhevsky, V. V. (2002). Interaction of neutrons with nanoparticles, *Phys. At. Nucl.* **65**, p. 400.

Rauch, H., Lemmel, H., Baron, M., and Loidl, R. (2002). Measurement of a confinement induced neutron phase, *Nature* **417**, p. 630.

Ahluwalia, D. V. (2002). Interface of gravitational and quantum realms, *Mod. Phys. Lett. A* **17**, p. 1135.

Kalbermann, G. (2002). Wave packets falling under a mirror, *J. Phys. A* **35**, p. 9829.

Bondarenko, L. N., Geltenbort, P., Korobkina, E. I., Morozov, V. I., and Panin, Yu. N. (2002). Cooling and heating of ultracold neutrons during storage, *Phys. At. Nucl.* **65**, p. 11.

Meyerovich, A. E., and Ponomarev, I. V. (2002). Surface roughness and size effects in quantum films, *Phys. Rev. B* **65**, p. 155413.

Friedrich, H., Jacoby, G., and Meister, C. G. (2002). Quantum reflection by Casimir – van der Waals potential tails, *Phys. Rev. A* **65**, p. 032902.

Herdegen, A., and Wawrzycki, J. (2002). Is Einstein's equivalence principle valid for a quantum particle? *Phys. Rev. D* **66**, p. 044007.

Nesvizhevsky, V. V., Boerner, H. G., Gagarski, A. M., Petukhov, A. K., Petrov, G. A., Abele, H., Baessler, S., Divcovic, G., Ruess, F. J., Stoeferle, Th., Westphal, A., Strelkov, A. V., Protasov, K. V., and Voronin, A.Yu. (2003). Measurement of quantum states of neutrons in the Earth's gravitational field, *Phys. Rev. D* **67**, p. 102002.

Nesvizhevskii, V. V. (2003). Quantum states of neutrons in a gravitational field and the interaction of neutrons with nanoparticles, *Physics-Uspekhi* **46**, p. 93.

Bertolami, O., and Nunes, F. M. (2003). Ultracold neutrons, quantum effects of gravity and the weak equivalence principle, *Class. Quant. Grav.* **20**, p. L61.

Druzhinina, V., and DeKieviet, M. (2003). Experimental observation of quantum reflection far from threshold, *Phys. Rev. Lett.* **91**, p. 193202.

Vahala, K. J. (2003). Optical microcavities, *Nature* **424**, p. 839.

Chrissomalakos, C., and Sudarsky, D. (2003). On the geometrical character of gravitation, *Gen. Relat. Grav.* **35**, p. 605.

Khorrami, M., Alimohammadi, M., and Shariati, A. (2003). Spin 0 and spin 1/2 quantum relativistic particles in a constant gravitational field, *Annal. Phys.* **304**, p. 91.

Watson, P. J. C. (2003). Bouncing neutrons and the neutron centrifuge, *J. Phys. C* **29**, p. 1451.

Robinett, R. W. (2004). Quantum wave packet revivals, *Phys. Rep.* **392**, p. 1.

Strelkov, A. V. (2004). Neutron storage, *Physics-Uspekhi* **47**, p. 511.

Frank, A. I., and Nosov, V. G. (2004). Quantum effects in a one-dimensional magnetic gravitational trap for ultracold neutrons, *JETP Lett.* **79**, p. 313.

Wawrzycki, J. (2004). Equality of the inertial and the gravitational masses for a quantum particle, *Acta Phys. Pol. B* **35**, p. 613.

Frank, A., van Isacker, P., and Gomez-Camacho, J. (2004). Probing additional dimensions in the universe with neutron experiments, *Phys. Lett. B* **582**, p. 15.

Alimohammadi, M., and Vakili, B. (2004). Spin 0 and spin 1/2 particles in a constant scalar-curvature background, *Annal. Phys.* **310**, p. 95.

Zabow, G., Conroy, R. S., and Prentiss, M. G. (2004). Coherent matter-wave manipulation in the diabatic limit, *Phys. Rev. Lett.* **92**, p. 180404.

Bini, D., Cherubini, C., and Mashhoon, B. (2004). Vacuum C metric and the gravitational Stark effect, *Phys. Rev. D* **70**, p. 044020.

Bini, D., Cherubini, C., and Mashhoon, B. (2004). Spin, acceleration and gravity, *Class. Quant. Grav.* **21**, p. 3893.

Pasquini, T. A., Shin, Y., Sanner, C., Saba, M., Schirotzek, A., Pritchard D. E., and Ketterle, W. (2004). Quantum reflection from a solid surface at normal incidence, *Phys. Rev. Lett.* **93**, p. 223201.

Frank, A. I., and Nosov V. G. (2004). Quantum effects in a one-dimensional magnetic gravitational trap for ultracold neutrons, *JETP Lett.* **79**, p. 313.

Nesvizhevsky, V. V., Petukhov, A. K., Boerner, H. G., Baranova, T. A., Gagarski, A. M., Petrov, G. A., Protasov, K. V., Voronin, A. Yu., Baessler, S., Abele, H., Westphal, A., and Lucovak, L. (2005). Study of the neutron quantum states in the gravity field, *Europ. Phys. J. C* **40**, p. 479.

Bertolami, O., Rosa, J. G., Aragao, C. M. L., Castorina, P., and Zappala, D. (2005). Noncommutative gravitational quantum well, *Phys. Rev. D* **72**, p. 025010.

Oberst, H., Tashiro, Y., Shimizu, K., and Shimizu, F. (2005). Quantum reflection of He-* silicon, *Phys. Rev. A* **71**, p. 052901.

Berberan-Santos, M., Bodunov, E., and Pogliani, L. (2005). Classical and quantum study of the motion of a particle in a gravitational field, *J. Math. Chem.* **37**, p. 101.

Belloni, M., Doncheski, M. A., and Robinett, M. W. (2005). Zero-curvature solutions of the one-dimensional Schrodinger equation, *Phys. Scrip.* **72**, p. 122.

Voronin, A. Yu., and Froelich, P. (2005). Quantum reflection of ultracold anti-hydrogen from a solid surface, *J. Phys. B* **38**, p. L301.

Voronin, A. Yu., Froelich, P., and Zygelman, B. (2005). Interaction of ultracold antihydrogen with a conducting wall, *Phys. Rev. A* **72**, p. 062903.

Kiefer, C., and Weber, C. (2005). On the interaction of mesoscopic quantum sytems with gravity, *Annal. Phys.* **14**, p. 253.

Leclerc, M. (2005). Mathisson-Papepetrou equations in metric and gauge theories of gravity in a Langrangian, *Class. Quant Grav.* **22**, p. 3203.

Ahluwalia-Khalilova, D. V. (2005). Minimal spatio-temporal extent of events, neutrinos, and the cosmological constant problem, *Int. J. Mod. Phys. D* **14**, p. 2151.

Nesvizhevsky, V. V., and Protasov, K. V. (2006). *Quantum States of Neutrons in the Earth's Gravitational Field: State of the Art in Trends in Quantum Gravity Research*, NOVA Scie. Publ., New York.

Voronin, A. Yu., Abele, H., Baessler, S., Nesvizhevsky, V. V., Petukhov, A. K., Protasov, K. V., and Westphal, A. (2006). Quantum motion of a neutron in a waveguide in the gravitational field, *Phys. Rev. D* **73**, p. 044029.

Cheng, Y. Y., and Meyerovich, A. E. (2006). Mode coupling in quantized high quality films, *Phys. Rev. B* **73**, p. 085404.

Pasquini, T. A., Saba, M., Jo, G. B., Shin, Y., Ketterle, W., Pritchard D. E., Savas, T. A., and Mulders, N. (2006). Low velocity quantum reflection of Bose-Einstein condensates, *Phys. Rev. Lett.* **97**, p. 093201.

Bertolami, O., and Rosa, J. G. (2006). Quantum and classical divide: the gravitational case, *Phys. Lett. B* **633**, p. 111.

Bertolami, O., Rosa, J. G., Aragao, C. M. L., Castorina, P., and Zappala, D. (2006). Scaling of variables and the relation between noncommutative parameters in noncommutative quantum mechanics, *Mod. Phys. Lett. A* **21**, p. 795.

Mather, W. H., and Fox, R. F. (2006). Coherent-state analysis of the quantum bouncing ball, *Phys. Rev. A* **73**, p. 032109.

Yoder, G. (2006). Using classical probability functions to illuminate the relation between classical and quantum physics, *Am. J. Phys.* **74**, p. 404.

Meyerovich, A. E., and Nesvizhevsky, V. V. (2006). Gravitational quantum states of neutrons in a rough waveguide, *Phys. Rev. A* **73**, p. 063616.

Arminjon, M. (2006). Dirac equation from the Hamiltonian and the case with a gravitational field, *Found. Phys. Lett.* **19**, p. 225.

Brau, F., and Buisseret, F. (2006). Minimal length uncertainty relation and gravitational quantum well, *Phys. Rev. D* **74**, p. 036002.

Banerjee, R., Roy, B. D., and Samanta, S. (2006). Remarks on the noncommutative gravitational quantum well, *Phys. Rev. D* **74**, p. 045015.

Arminjon, M. (2006). Post-Newtonian equation for the energy levels of a Dirac particle in a static metric, *Phys. Rev. D* **74**, p. 065017.

Witthaut, D., and Korsch, S. J. (2006). Uniform semiclassical approximations of the nonlinear Schrödinger equation by a Painleve mapping, *J. Phys. A* **39**, p. 14687.

Boulanger, N., Spindel, Ph., and Buisseret, F. (2006). Bound states of Dirac particles in gravitational fields, *Phys. Rev. D* **74**, p. 125014.

Wu, N. (2006). Non-relativistic limit of Dirac equations in gravitational field and quantum effects of gravity, *Comm. Theor. Phys.* **45**, p. 452.

Baessler, S., Nesvizhevsky, V. V., Protasov, K. V., and Voronin, A.Yu. (2007). Constraint on the coupling of axionlike particles to matter via an ultracold neutron gravitational spectrometer, *Phys. Rev. D* **75**, p. 075006.

Brax, Ph., van de Bruck, C., Davis, A. C., Mota, D. F., and Shaw, D. (2007). Detecting chameleons through Casimir force measurements, *Phys. Rev. D* **76**, p. 124034.

Accioly, A., and Blas, H. (2007). Metric-scalar gravity with torsion and the mesurability of the non-minimal coupling, *Mod. Phys. Let. A* **22**, p. 961.

Schmidt-Wellenburg, Ph., Barnard, J., Geltenbort, P., Nesvizhevsky, V. V., Plonca, C., Soldner, T., and Zimmer, O. (2002). Efficient extraction of a collimated ultra-cold neutron beam using diffusive channels, *Nucl. Instr. Meth. A* **577**, p. 623.

Adhikari, R., Cheng, Y., Meyerovich, A. E., and Nesvizhevsky, V. V. (2007). Constraint on the coupling of axionlike particles to matter via an ultracold neutron gravitational spectrometer, *Phys. Rev. A* **75**, p. 063613.

Buisseret, F., Silvestre-Brac, B., and Mathieu, V. (2007). Modified Newton's law, braneworlds, and the gravitational quantum well, *Class. Quant. Grav.* **24**, p. 855.

Pignol, G., Protasov, K. V., and Nesvizhevsky, V. V. (2007). A note on spontaneous emission of gravitons by a quantum bouncer, *Class. Quant. Grav.* **24**, p. 2439.

Saha, A. (2007). Time-space non-commutativity in gravitational quantum well scenario, *Europ. Phys. J. C* **51**, p. 199.

Nesvizhevsky, V. V., Pignol, G., Protasov, K. V., Quemener, G., Forest, D., Ganau, P., Mackowski, J. M., Michel, C., Montorio, J.L., Morgado, M., Pinard, L., and Remillieux, A. (2007). Comparison of specularly reflecting mirrors for GRANIT, *Nucl. Instr. Meth. A* **578**, p. 435.

Romera, E., and de los Santos, F. (2007). Identifying wave-packet fractional revivals by means of information entropy, *Phys. Rev. Lett.* **99**, p. 263601.

Silenko, A. J., and Teryaev, O. V. (2007). Equivalence principle and experimental tests of gravitational spin effects, *Phys. Rev. D* **76**, p. 061101.

Mann, R. B., and Young, M. B. (2007). Perturbative quantum gravity coupled to particles in (1+1) dimension, *Class. Quant. Grav.* **24**, p. 951.

Nesvizhevsky, V. V., Pignol, G., and Protasov, K. V. (2008). Neutron scattering and extra-short-range interactions, *Phys. Rev. D* **77**, p. 034020.

Nesvizhevsky, V. V., Plonka-Spehr, P., Protasov, K., Schreckenbach, K., Soldner, T., and Zimmer, O. (2008). Particle physics with slow neutrons. Proceedings of the International Workshop on Particle Physics with Slow Neutrons, Institut Laue-Langevin, Grenoble, France, May 29–31, 2008. Preface. *Nucl. Instr. Meth. A* **611**, p. VII.

Gonzalez, G. (2008). Quantum bouncer with quadratic dissipation, *Rev. Mex. Fis.* **54**, p. 5.

Nesvizhevsky, V. V., Petukhov, A. K., Protasov, K. V., and Voronin, A. Yu. (2008). Centrifugal quantum states of neutrons, *Phys. Rev. A* **78**, p. 033616.

Longhi, S. (2008). Quantum bouncing ball on a lattice: An optical realization, *Phys. Rev. A* **77**, p. 035802.

Barnard, J., and Nesvizhevsky, V. V. (2008). Analysis of a method for extracting angularly collimated UCNs from a volume without losing the density inside, *Nucl. Instr. Meth. A* **591**, p. 431.

Bastos, C., and Bertolami, O. (2008). Berry phase in the gravitational quantum well and the Seiberg-Witten map, *Phys. Lett. A* **372**, p. 5556.

Giri, P. R., and Roy, P. (2008). The non-commutative oscillator, symmetry and the Landau problem, *Europ. Phys. J. C* **57**, p. 835.

Arminjon, M. (2008). Dirac-type equations in a gravitational field, *Found. Phys.* **38**, p. 1020.

Arminjon, M. (2008). Main effects of the Earth's rotation on the stationarry states of ultra-cold neutrons, *Phys. Lett. A* **372**, p. 2196.

Belloni, M. (2009). Constraints on Airy function zeros from quantum-mechanical sum rules, *J. Phys. A* **42**, p. 075203.

Kreuz, M., Nesvizhevsky, V. V., Schmidt-Wellenburg, Ph., Soldner, T., Thomas, M., Boerner, H. G., Naraghi, F., Pignol, G., Protasov, K. V., Rebreyend, D., Vezzu, F., Flaminio, R., Michel, C., Morgado., N., Pinard, L., Baessler, S., Gagarski, A. M., Grigorieva, L. A., Kuzmina, T. M., Meyerovich, A. E., Mezhov-Deglin, L. P., Petrov, G. A., Strelkov, A. V., and Voronin, A. Yu. (2009). A method to measure the resonance transitions between the gravitationally bound quantum states of neurons in the GRANIT spectrometer, *Nucl. Instr. Meth. A* **611**, p. 326.

Sanuki, T., Komamiya, S., Kawasaki, S., and Sonoda, S. (2009). Proposal for measuring the quantum states of neutrons in the gravitational field with a CCD-based pixel sensor, *Nucl. Instr. Meth. A* **600**, p. 657.

Jakubek, J., Schmidt-Wellenburg, Ph., Geltenbort, P., Platkevic, M., Plonka-Spehr, C., Solc, J., and Soldner, T. (2009). A coated pixel device TimePix with micron spatial resolution for UCN detection, *Nucl. Instr. Meth. A* **600**, p. 651.

Ernest, A. D. (2009). Gravitational eigenstates in weak gravity: I. Dipole decay rates of charged particles, *J. Phys. A* **42**, p. 115207.

Ernest, A. D. (2009). Gravitational eigenstates in weak gravity: II. Further approximate methods for decay rates, *J. Phys. A* **42**, p. 115208.

Della Valle, G., Savoini, M., Ornigotti, M., Laporta, P., Foglietti, V., Finazzi, M., Duo, L., and Longhi, S. (2009). Experimental observation of a photon bouncing ball, *Phys. Rev. Lett.* **102**, p. 180402.

Jakubek, J., Platkevic, M., Schmidt-Wellenburg, Ph., Geltenbort, P., Plonka-Spehr, C., and Daum, M. (2009). Position-sensitive spectroscopy of ultra-cold neutrons with TimePix pixel detector, *Nucl. Instr. Meth. A* **607**, p. 45.

Baessler, S. (2009). Gravitationally bound quantum states of ultracold neutrons and their applications, *J. Phys. G* **36**, p. 104005.

Abele, H., Jenke, T., Stadler, T., and Geltenbort, P. (2009). QuBounce: the dynamics of ultra-cold neutrons falling in the gravity potential of the Earth, *Nucl. Phys. A* **827**, p. 593C.

Cubitt, R., Nesvizhevsky, V. V., Petukhov, A. K, Voronin, A. Yu., Pignol, G., Protasov, K. V., and Gurshijants, P. (2009). Methods of observation of the centrifugal quantum states of neutrons, *Nucl. Instr. Meth. A* **611**, p. 322.

Jenke, T., Stadler, T., Abele, H., and Geltenbort, P. (2009). Q-Bounce experiments with quantum bouncing ultracold neutrons, *Nucl. Instr. Meth. A* **611**, p. 318.

Morozova, V. S., and Ahmedov, B. J. (2009). Quantum interference effects in slowly rotating nut space-time, *Int. J. Mod. Phys. D* **18**, p. 107.

Cronin, A. D., Shmiedmayer, J., and Pritchard, D. E. (2009). Optics and interferometry with atoms and molecules, *Rev. Mod. Phys.* **81**, p. 1051.

Schmidt-Wellenburg, Ph., Andersen, K. H., Courtois, P., Kreuz, M., Mironov, S., Nesviwhevsky, V. V., Pignol, G., Protasov, K. V., Soldner, T., Vezzu, F. and Zimmer, O. (2009). Ultracold-neutron infrastructure for the gravitational spectrometer GRANIT, *Nucl. Instr. Meth. A* **611**, p. 267.

Nesvizhevsky, V. V. (2010). Near-surface quantum states of neutrons in the gravitational and centrifugal potentials, *Physics-Uspekhi* **53**, p. 645.

Nesvizhevsky, V. V., Voronin, A. Yu., Cubitt, R., and Protasov, K. V. (2010). Neutron whispering gallery, *Nature Phys.* **6**, p. 114.

Rauch, H. (2010). Quantum optics: Neutrons in a whispering gallery, *Nature Phys.* **6**, p. 79.

Nesvizhevsky, V. V., Cubitt, R., Protasov, K. V., and Voronin, A. Yu. (2010). The whispering gallery effect in neutron scattering, *New J. Phys.* **12**, p. 113050.

Baessler, S., Nesvizhevsky, V. V., Pignol, G., Protasov, K. V., and Voronin, A. Yu. (2010). Constraints on spin-dependent interactions using gravitational quantum levels of ultracold neutrons, *Nucl. Instr. Meth. A* **611**, p. 149.

de los Santos, F., Guglien, C., and Romera, E. (2010). Application of new uncertainty relations to wave packet revivals, *Phys. E* **42**, p. 303.

Robinett, R. W. (2010). The Stark effect in linear potentials, *Europ. J. Phys. A* **31**, p. 1.

Abele, H., Jenke, T., Leeb, H., and Schmiedmayer, J. (2010). Ramsey's method of separated oscillating fields and its application to gravitationally induced quantum phase shifts, *Phys. Rev. D* **81**, p. 065019.

Kawasaki, S., Ichikawa, G., Hino, M., Kamiya, Y., Kitaguchi, M., Komamiya, S., Sanuki, T., and Sonoda, S. (2010). Development of a pixel detector for ultra-cold neutrons, *Nucl. Instr. Meth. A* **615**, p. 42.

Zhao, B. S., Schewe, H. C., Meijer, G., and Schollkopf, W. (2010). Coherent reflection of helium atom beams from rough surfaces at grazing incidence, *Phys. Rev. Lett.* **105**, p. 133203.

Saha, A. (2010). Galilean symmetry in a noncommutative gravitational quantum well, *Phys. Rev. D* **81**, p. 125002.

Ayorinde, O. A., Chisholm, K., Belloni, M., and Robinett, R. W. (2010). New identities from quantum-mechanics sum rules of parity-related potentials, *J. Phys. A* **43**, p. 235202.

Nakimov, A., Turimov, B., Abdujabbarov, A., and Ahmedov, B. (2010). Quantum interference effects in Horava–Lifshitz gravity, *Mod. Phys. Lett. A* **25**, p. 3115.

Contreras-Reyes, A. M., Guerout, R., Neto, P. A. M., Dalvit, D. A. R., Lambrecht, A., and Reynaud, S. (2010). Casimir–Polder interaction between an atom and a dielectric grating, *Phys. Rev. A* **82**, p. 052517.

Kajari, E., Harshman, N. L., Rasel, E. M., Stenholm, M., Sussmann, G., and Schleich, W. P. (2010). Inertial and gravitational mass in quantum mechanics, *Appl. Phys. B* **100**, p. 43.

Adler, R. J., and Chen, P. (2010). Gravitomagnetism in quantum mechanics, *Phys. Rev. D* **82**, p. 025004.

Alves, A., and Bertolami, O. (2010). Unparticle inspired corrections to the gravitational quantum well, *Phys. Rev. D* **82**, p. 047501.

Castello-Branco, K. H. C., and Martins A. G. (2010). Free-fall in a uniform gravitational field in noncommutative quantum mechanics, *J. Math. Phys.* **51**, p. 102106.

Nozari, K., and Pedram, P. (2010). Minimal length and bouncing-particle spectrum, *Eurp. Phys. Lett.* **92**, p. 50013.

Mueller, H., Peters, A., and Chu, S. (2010). A precision measurement of the gravitational redshift by the interference of matter waves, *Nature* **463**, p. 926.

Wolf, P., Blanchet, L., Borde, C. J., Reynaud, S., Salomon, C., and Cohen-Tannoudji, C. (2011). Atom gravimeters and gravitational redshift, *Nature* **267**, p. E1.

Mueller, H., Peters, A., and Chu, S. (2010). Atom gravimeters and gravitational redshift. Reply, *Nature* **467**, p. E2.

Voronin, A. Yu., Froelich, P., and Nesvizhevsky, V. V. (2011). Gravitational quantum states of antihydrogen, *Phys. Rev. A* **83**, p. 032903.

Baessler, S., Beau, M., Kreuz, M., Kurlov, V. N., Nesvizhevsky, V. V., Pignol, G., Protasov, K. V., Vezzu, F., and Voronin A. Yu. (2011). The GRANIT spectrometer, *Compt. Rend. Phys.* **12**, p. 707.

Baessler, S., Gagarski, A. M., Lychagin, E. V., Mietke, A., Muzychka, A. Yu., Nesvizhevsky, V. V., Pignol, G., Strelkov, A. V., Toperverg, B. P., and Zhernenkov, K. (2011). New methodical developments for GRANIT, *Compt. Rend. Phys.* **12**, p. 729.

Antoniadis, I., Baessler, S., Bertolami, O., Dubbers, D., Meyerovich, A. E., Nesvizhevsky, V. V., Protasov, K. V., and Reynaud, S. (2011). Workshop GRANIT-2010, 14–19 February 2010, Les Houches, France. *Compt. Rend. Phys.* **12**, p. 703.

Antoniadis, I., Baessler, S., Buchner, M., Fedorov, V. V., Hoedl, S., Lambrecht, A., Nesvizhevsky, V. V., Pignol, G., Protasov, K. V., Reynaud, S., and Sobolev, Yu. (2011). Short-range fundamental forces, *Compt. Rend. Phys.* **12**, p. 755.

Judd, T. E., Scott, R. G., Martin, A. M., Kaszmarek, B., and Fromhold, T. M. (2011). Quantum reflection of ultracold atoms from thin films, graphene and semiconductor, *New J. Phys.* **13**, p. 083020.

Chang, L. N., Levis, Z., Minic, D., and Takeuchi, T. (2011). On the minimal length uncertainty relation and the foundations of string theory, *Adv. High Ener. Phys.* **493514**, p. 1.

Floratos, E. G., Leontaris, G. K., and Vlachos, N.D. (2011). Gravitational atom in compactified extra dimensions, *Phys. Lett. B* **694**, p. 410.

Kobakhidze, A. (2011). Gravity is not an entropic force, *Phys. Rev. D* **83**, p. 021502.

Pedram, P., Nozari, K., and Taheri, S. H. (2011). The effects of minimal length and maximal momentum on the transition rate of ultracold neutrons in gravitational field, *J. High Ener. Phys.* **3**, p. 093.

Escobas, M., and Meyerovich, A. E. (2011). Beams of gravitationally bound ultracold neutrons in rough waveguides, *Phys. Rev. A* **83**, p. 033618.

Wolf, P., Blanchet, L., Borde, C. J., Reynaud, S., Salomon, C., and Cohen-Tannoudji, C. (2011). Does an atomic interferometer test the gravitational redshift at the Compton frequency? *Class. Quant. Grav.* **28**, p. 145017.

Jenke, T., Geltenbort, P., Lemmel, H., and Abele, H. (2011). Realization of a gravity-resonance-spectroscopy technique, *Nature Phys.* **7**, p. 468.

Bastos, C., Bertolami, O., Dias, N. C., and Prata, J. N. (2011). Entropic gravity, phase-space noncommutativity and the equivalence principle, *Class Quant. Grav.* **28**, p. 125007.

Durstberger-Rennhofer, K., Jenke, T., and Abele, H. (2011). Probing the neutron's electric neutrality with Ramsey spectroscopy of gravitational quantum states of ultracold neutrons, *Phys. Rev. D* **84**, p. 036004.

Chaichian, M., Oksanen, M., and Tureanu, A. (2011). On gravity as an entropic force, *Phys. Lett. B* **702**, p. 419.

Andresen, G. B. *et al* (ALPHA)(2011). *Nature Phys.* **7**, p. 558.

Brax, Ph., and Pignol, G. (2011). Strongly coupled chameleons and the neutronic quantum bouncer, *Phys. Rev. Lett.* **107**, p. 111301.

Bourdel, T., Doser, M., Ernest, A. D., Voronin, A. Yu., and Voronin, V. V. (2011). Quantum phenomena in gravitational field, *Compt. Rend. Phys.* **12**, p. 779.

Romera, E. (2011). Revivals of zitterbewegung of a bound localized Dirac particle, *Phys. Rev. A* **84**, p. 052102.

Hasegawa, Y., and Rauch, H. (2011). Quantum phenomena explored with neutrons, *New J. Phys.* **13**, p. 115010.

Gonzalez, G., and Piza, H. I. (2011). WKB quantization for completely bound quadratic dissipative systems, *Rev. Mex. Fis.* **57**, p. 481.

Liu, W., Neshev, D. N., Miroshnechenko, A. E., Shadrivov, I. D., and Kivshar Yu. S. (2011). Bouncing plasmonic waves in half-parabolic potentials, *Phys. Rev. A* **84**, p. 063805.

Amole, C. *et al* (ALPHA) (2012). Resonant quantum transitions in trapped antihydrogen atoms, *Nature* **483**, p. 439.

Voronin A. Yu., Nesvizhevsky, V. V., and Reynaud, S. (2012). Whispering-gallery states of antihydrogen near a curved surface, *Phys. Rev. A* **85**, p. 014902.

Nesvizhevsky, V. V., Voronin, A. Yu., Cubitt, R., and Protasov, K. V. (2012). A note on observation and theoretical description of the neutron whispering gallery effect, *J. Phys. Conf. Ser.* **340**, p. 012020.

Adler, R. J., Chen, P., and Varani, E. (2012). Gravitomagnetism and spinor quantum mechanics, *Phys. Rev. D* **85**, p. 025016.

Konno, K., and Takahachi, R. (2012). Spacetime rotation-induced Landau quantization, *Phys. Rev. D* **85**, p. 061502.

Pedram, P. (2012). Minimal length and the quantum bouncer: a nonperturbative study, *Int. J. Theor. Phys.* **51**, p. 1901.

Ernest, A. D., and Collins, M. P. (2012). Structural features of high-n gravitational eigenstates, *Grav. Cosm.* **18**, p. 242.

Chaichian, M., Oksanen, M., and Tureanu, A. (2012). On entropic gravity: the entropy postulate, entropy content of screens and relaion to quantum mechanics, *Phys. Lett. B* **712**, p. 272.

Codau, C., Nesvizhevsky, V. V., Fertl, M., Pignol, G., and Protasov, K. V. (2012). Transitions between levels of a quantum bouncer by a noise-like perturbation, *Nucl. Instr. Meth. A* **677**, p. 10.

Martin, J. (2012). Everything you always wanted to know about the cosmological constant problem (but were afraid to ask), *Compt. Rend. Phys.* **13**, p. 566.

Perez, P., and Sacquin, Y. (2012). The GBAR experiment: gravitational behaviour of antihydrogen at rest, *Class. Quant. Grav.* **29**, p. 184008.

Ruckle, L. J., Belloni, M., and Robinett, R. W. (2012). The biharmonic oscillator and asymmetric linear potentials: from classical trajectories to momentum-space probability densities in the extreme quantum limit, *Europ. J. Phys.* **33**, p. 1505.

Debu, P. (2012). GBAR: gravitational behaviour of antihydrogen at rest, *Hyper. Inter.* **212**, p. 51.

Perez, P., and Sacquin, Y. (2012). The GBAR experiment: gravitational behaviour of antihydrogen at rest. *Class. Quant. Grav.* **29**, p. 184008.

Doser, M. *et al* (AEgIS Collaboration) (2012). Exploring the WEP with a pulsed cold beam of antihydrogen. *Clas. Quant. Grav.* **29**, p. 184009.

Greenberger, D. M., Schleich, W. P., and Rasel, E. M. (2012). Relativistic effects in atom and neutron interferometry and the differences between them, *Phys. Rev. A* **86**, p. 063622.

Beringer, J. *et al.* (Particle Data Group) (2012). *Phys. Rev. D* **86**, p. 010001.

Amelino-Camelia, G. (2013). Quantum-spacetime phenomenology, *Liv. Rev. Rel.* **16**, p. 5.

Brax, Ph., Pignol, G., and Roulier, D. (2013). Probing strongly couples chameleons with slow neutrons, *Phys. Rev. D* **88**, p. 083004.

Onofrio, R. (2013). On weak interactions as short-distance manifestations of gravity, *Mod. Phys. Lett. A* **28**, p. 1350022.

Bonder, Y., Fischbach, E., Hernandez-Coronado, H., Krause, D. E., Rohrbach, Z., and Sudarsky, D. (2013). Testing the equivalence principle with unstable particles, *Phys. Rev. D* **87**, p. 125021.

Pedram, P. (2013). Exact ultra cold neutrons' energy spectrum in gravitational quantum mechanics, *Europ. Phys. J. C* **73**, p. 2609.

Romera, E., and de los Santos, F. (2013). Fisher information, nonclassicality and quantum revivals, *Phys. Lett. A* **377**, p. 2284.

Abreu, E. M. C., and Neto, J. A. (2013). Considerations on gravity as an entropic force and entangled states, *Phys. Lett. B* **727**, p. 524.

Dufour, G., Gerardin, A., Guerout, A., Lambrecht, A., Nesvizhevsky, V. V., Reynaud, S., and Voronin A. Yu. (2013). Quantum reflection of antihydrogen from the Casimir potential above material slabs, *Phys. Rev. A* **87**, p. 012901.

Dufour, G., Guerout, A., Lambrecht, A., Nesvizhevsky, V. V., Reynaud, S., and Voronin A. Yu. (2013). Quantum reflection of antihydrogen from nanoporous media, *Phys. Rev. A* **87**, p. 033506.

Garcia, T., de los Santos, F., and Romera, E. (2014). Fisher–Shannon product and quantum revivals in wavepacket dynamics, *Phys. A* **394**, p. 394.

Dufour, G., Debu, P., Lambrecht, A., Nesvizhevsky, V. V., Reynaud, S., and Voronin A. Yu. (2014). Shaping the distribution of vertical velocities of antihydrogen in GBAR, *Europ. Phys. J. C* **74**, p. 2731.

Index

www.ingramcontent.com/pod-product-compliance
Lightning Source LLC
Chambersburg PA
CBHW060237220326
41598CB00027B/3967